50
Daily Telegraph
Brain-Twisters

By the same author:

FIFTY OBSERVER BRAIN-TWISTERS
ONE HUNDRED BRAIN-TWISTERS
ANATOMY OF THE CROSSWORD
ADVENTURES IN MATHEMATICS
IT'S ALL DONE BY NUMBERS
FIGURE IT OUT

50
Daily Telegraph
BRAIN-TWISTERS

*A book of mathematical puzzles
and reasoning problems*

D. StP. BARNARD

With illustrations by the author

JAVELIN BOOKS
POOLE · DORSET

First published in the UK 1985 by Javelin Books,
Link House, West Street, Poole, Dorset, BH15 1LL

Copyright © 1985 D. StP. Barnard

Distributed in the United States by
Sterling Publishing Co., Inc.,
2 Park Avenue, New York, NY 10016

British Library Cataloguing in Publication Data

Barnard, D. St. P.
 50 Daily telegraph brain-twisters : a book
 of mathematical puzzles and reasoning problems.
 1. Mathematical recreations
 I. Title II. The Daily Telegraph
 793.7'4   QA95

ISBN 0 7137 1612 6

All rights reserved. No part of this book may be
reproduced or transmitted in any form or by any
means, electronic or mechanical, including photocopying,
recording or any information storage and retrieval system,
without permission in writing from the publisher.

This book is sold subject to the conditions that it shall
not, by way of trade or otherwise, be lent, re-sold, hired
out or otherwise circulated without the publisher's prior
consent in any form of binding or cover other than that in
which it is published and without a similar condition
including this condition being imposed on the subsequent
purchaser.

Typeset by Latimer Trend & Company Ltd,
Plymouth, Devon.

Printed in Great Britain by
Cox & Wyman Ltd., Reading, Berks.

CONTENTS

THE PROBLEMS   8—76

Food for Thought

1	A Little Choosey	12
2	Poulterer's Dozen	13
3	Mix Up	14
4	Measure for Measure	15
5	Well Done	16
6	A Piece of Cake	17

Going Places

7	Far Enough	20
8	Local Colour	21
9	Mystery Tour	22
10	Air Miles	24
11	Touranian Tours	25
12	Heare and Thare	26
13	World Tour	27

It's on the Cards

14	Red or Black	30
15	Think of Two Cards	31
16	Straight Answer Needed	32
17	Handy Question	33
18	Win or Lose	34
19	Pick-a-Pack	35
20	Very Tricky Question	36

Who's Who

21	What's in a Name?	38
22	French and English	39
23	Tabulation	40
24	Smoko	41
25	Sorting out the Mess	42
26	Gooseberry Fool	43

Fun and Games

27	Catch Patience	46
28	Odds and Evens	47
29	Grand Chain	48
30	All Buttoned Up	49
31	Diamond Chain	51
32	Checkel	52
33	Break'n the Bank	53

On the Square

34	Take Your Pick	56
35	Square Pairs	57
36	Each Way Accumulator	58
37	Crossnumber	59
38	PSC Square	60

| 39 | Digi-Tally | 61 |

Mental Blockage

40	Exhibitionism	64
41	Cut It Out	65
42	Make It Up	66
43	In Black and White	67
44	Rubikitis	68
45	Symboliosis	69

Quid Est Veritas?

46	Royal Dilemma	72
47	Age Old Question	73
48	Amphibolia	74
49	But Which of Us?	75
50	Tell Me Truly	76

| LEADS | 77—99 |
| SOLUTIONS | 101—126 |

THE PROBLEMS

In preparing these problems for publication I have revised some of them where it seemed to me that an extra word of explanation here or there might clarify their presentation.

To provide some coherent structure to the book, the puzzles have been grouped together in eight sections, each dealing with a common theme such as food, travel, identity, etc, but within each section a wide variety of solving methods are called for.

As with most of my previous puzzle books, I have inserted a section of Leads, the purpose of which is to aid a solver who may find a problem too elusive in its original form, but who may be able to mount a renewed attack if given a hint or working diagram that will lead him along the right lines. In virtually every case, the reasoning given in the Solution is dependent upon that proposed in the Lead, and a solver who is intent on pursuing an exhaustive analysis should satisfy himself as to the points made in the Lead before proceeding to the Solution, which in each case has been expanded well beyond the restricted space normally available in a newspaper column.

Both the Leads and the Solutions sections are prefaced with an explanation of how those sections may be most profitably used. An explanation of the mathematical and reasoning conventions used in analysing the puzzles is given in the introduction to the Leads on page 77.

I take this opportunity to acknowledge my deep indebtedness to Peter Mabey who, for many years, has acted as 'invigilator' to my journalistic puzzle columns. His timely interventions at draft stage have saved me from perpetrating many an horrendous howler. He has also been responsible for suggesting a number of improvements in presentation and methods of solution, but if any errors have crept into this present re-edition of the puzzles, the fault is entirely mine.

I must also express my thanks to the proprietor and editor of the *Daily Telegraph* for permission to use the paper's name in the title of this collection of items from that paper's distinguished columns.

Finally, I must thank those thousands of solvers who, over the past thirty-odd years, have taught me far more about puzzles than ever I have been able to teach them.

D. StP. B.
Cheltenham Spa 1985

FOOD FOR THOUGHT

*A few preliminary items of food to provide
light sustenance for your journey through the tortuous
labyrinths of thought that lie ahead.*

1 A LITTLE CHOOSEY

Mrs Chewsey has quite a problem with her little Chewseys when it comes to chewing.

Five of the brats choose not to chew cabbage, five will not chew cauliflower, and five eschew carrots.

Of those who chew cabbage, only four choose to chew carrots. Of those who chew carrots, only three will chew cauliflower. And of those who will chew cauliflower, only two will chew also cabbage.

Only one of the little Chewseys is not at all choosey, and is happy to chew anything.

How many little Chewseys are there in the family, and how many choose to chew what?

Lead – page 78
Solution – page 101

2 POULTERER'S DOZEN

The pavement outside our local grocery shop is littered with scraps of paper bearing hastily scribbled calculations, despairingly abandoned by would-be shoppers. You see, our grocer is a compulsive gimmick-monger, who is forever thinking up new ways of tempting his customers into trying their luck.

A few years ago, when halfpennies were still in existence, there appeared a notice in his window proclaiming:

SPECIAL OFFER
FIRST QUALITY EGGS
At this week's special price, fifty pence can buy you
as many eggs as the number of pence
you would receive in change if you were to proffer a
fifty-pence
piece for a dozen eggs.
A FREE EGG-CUP will be given to any customer
who buys thirteen eggs and proffers
the exact amount of money for them.

I was just pondering the matter when young Quaddle (who is too clever by half for my liking) strutted pompously out of the shop with his thirteen eggs and his free egg-cup.

To ask him how much he paid for the wretched things would be below my dignity, so I am still trying to work out what money I should have offered the grocer to be able to claim a free egg-cup of my own.

Have you any idea?

Lead – page 78
Solution – page 102

3 MIX UP

'Just measure out the ingredients for the oat-cake while I slip down to the corner shop for some almonds to put on the top,' said my wife. 'You'll find the recipe on page 99 of Mrs Waspton's.'

I rather like oat-cake, so I looked up the recipe:

Flour	9 oz
Oatmeal	10 oz
Butter	8 oz
Sugar	6 oz
Eggs (large)	6

But somehow I got a little confused. What happened was that I put into the bowl 10 oz of flour with 9 oz of oatmeal, instead of the other way around – and 6 oz of butter with 8 oz of sugar, instead of vice versa.

And I mixed them – oh yes, I mixed them all too well!

Now it is just as impossible to unmix well-mixed ingredients as it is to unscramble eggs, so I am wondering if I can add a little of this or that to bring the mixture up to its proper oak-cake consistency. And since the cake is going to be bigger than intended anyway, I want to know the smallest quantity of whatever has to be added to achieve this noble aim.

And I want to know urgently – if possible, before the wife comes back from the shop and discovers what an idiot I am.

Lead – page 79
Solution – page 102

4 MEASURE FOR MEASURE

How can Tom divide 10 pints of milk into two equal portions, given only a 1-gallon bowl, a jug that holds exactly....

No, no! I really must not invite groans and cat-calls from experienced puzzle-solvers, who are bored to death with being told the capacities of jugs, flasks and bowls, and are then asked to measure out a given quantity. So to avoid further boredom, I will tell you right away how Tom did it.

Since he knew that both the jug and the flask were just half full to begin with, he started by pouring all the contents of the jug into the empty 1-gallon bowl. He then filled the bowl to the brim with milk from the flask, and next filled the jug from the bowl. Finally he emptied the contents of the jug into the flask.

Hey presto! Ten pints of milk equally divided. Nothing to it, is there?

But what were the overall capacities of the three vessels?

Lead – page 79
Solution – page 103

5 WELL DONE

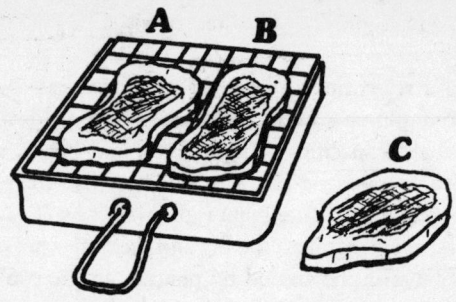

Fig. 1

Three juicy steaks, but (as Fig. 1 shows) a grill-pan large enough to hold only two at a time. That is the problem confronting me.

Now it takes no time at all to throw a steak on the pan and pop it under the grill – or to turn it, or take it off for that matter. But it does take one minute to prepare and season *each* side of each steak *before* exposing that side to the heat.

I like my steak well done – for a total of four minutes on each side. My wife prefers medium – a total of three minutes per side for her. And her mother likes her steak cooked for only two minutes on each side – a rare one is mother-in-law!

It is now exactly 6 o'clock. What is the earliest possible time by which I can have all three steaks done to order?

Lead – page 80
Solution – page 103

6 A PIECE OF CAKE

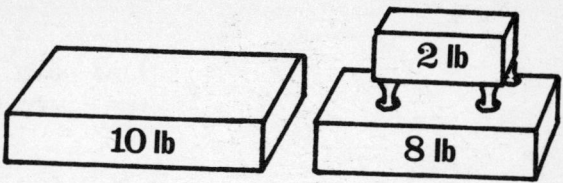

Fig. 2

Finally, to complete this gastronomic course – a piece of cake! Or is it all that easy?

When I first published this problem, I imagined that I had found an ingenious answer, but when my solution appeared in the newspaper, nine readers wrote in to describe a method which I thought far neater than the one I had suggested.

The problem was about my visit to the pastrycook's for the purpose of ordering a cake for my daughter's wedding.

'A square two-tier cake?' he echoed. 'I'm sorry, but the only square wedding cake I have left is a single-tier cake of 10 pounds. There is no time to bake another for next week; they do take time to mature, you know.'

'But Pamela insists on a cake with an 8-pound base, and a 2-pound tier of the same thickness,' I explained.

'Odd,' mused the cook, 'that comes to 10 pounds altogether, doesn't it? Wait a minute, I have an idea! Maybe I could cut up the cake I have, and fit the pieces together to make what you want. Once it is iced, no one will notice the joins, though to minimise crumbling I must try to do it with the fewest possible cuts.'

Fig. 2 shows you what the pastrycook had in stock, and what I wanted. How could the pastrycook have best cut his cake to fulfil my order?

If you think you have a good practical answer, do consult the Lead before checking your suggestion in the Solution section. The crumbs of information you find there may cause you to chew over your 'slice' of cake a little more before swallowing it.

Lead – page 80
Solution – page 104

GOING PLACES

Puzzles which take us further afield, and lead ultimately all over the world.

7 FAR ENOUGH

The road from Canostra was hot and dusty, and I was beginning to wish that I hadn't bothered about visiting the tiny, remote village of Lopez with its church hewn from solid rock.

'How far have we come?' I asked my Priscillian guide, as we rounded yet one more bend in the road that snaked over the barren hills.

'Half as far as we still have to go,' he replied gruffly.

We trudged on another half mile.

'How far have we still to go?' I asked wearily.

'Half as far again as we have already come,' was the unhelpful reply, which left me wondering whether or not we were heading in the wrong direction.

In fact we were not, but can you tell me how far it is from Canostra to Lopez?

Lead – page 81
Solution – page 104

8 LOCAL COLOUR

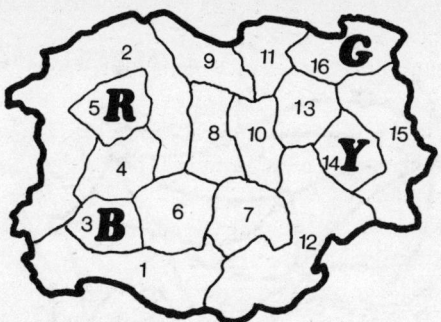

Fig. 3

Included in the brochures published by the Problemanian Tourist Bureau is a small sketch map showing the island's sixteen provinces, identified in Fig. 3 by the numbers 1–16.

For the sake of clarity, I decided to colour the provinces in red, blue, green, and yellow. The map in Fig. 3 (lettered R,B,G,Y) shows my attempt so far at colouring in the provinces with my felt-tipped pen.

To avoid confusion, I do not want any two contiguous provinces to share the same colour, and I fear that if I just proceed haphazardly I could well run myself into a corner that would require a fifth colour – which I haven't got.

I have heard that, by recourse to a lengthy computer program, it has now been proved that no map drawn on a plane surface (or a sphere for that matter) needs more than four colours to shade it in the way I have described, but that doesn't seem to help much when confronted by a real situation.

Carrying on from where I left off, can you shade the remaining districts with the four colours in such a way that no contiguous districts share the same colour?

Lead – page 81
Solution – page 105

9 MYSTERY TOUR

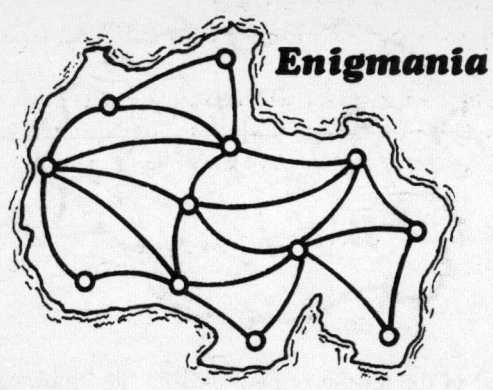

Fig. 4

'There are several round tours of the country,' explained the clerk at the Enigmanian Tourist Bureau. 'For instance, there is the Highland route: B,K,J,T,C,A,L,B, or the Lowland route: B,L,R,M,S,D,F,B. But the most popular tour this season is the Scenic route: B,J,C,A,R,D,B. Each of these tours visits the towns in the order given.'

He glanced furtively over his shoulder, and whispered, 'Since the Revolution all our towns have been given single-letter names, you know. It is part of our programme to eradicate reactionary sentiments.'

He handed me the map (Fig. 4). It showed the towns with their connecting roads.

'There are no names on the map,' I protested. 'I mean it hasn't even got the letters.'

'Revolutionary security regulations,' hissed the clerk. 'We are not allowed to divulge the individual location of each town.'

He lowered his voice to an even more conspiratorial level and added, 'But you can pencil in the names – I mean the letters – for yourself, provided you don't let anyone see you doing it.'

'Can you *tell* me what they are?'

'That would be more than my job's worth,' he replied.

Can *you* tell me?

(If at first you don't succeed, don't give up. After all, there are only 479,001,600 ways in which the twelve letters can be entered, and one of those ways is sure to be right.)

Lead – page 81
Solution – page 106

10 AIR MILES

Travellers are not helped greatly by tourist bureaux that issue out-of-date brochures. Amongst the literature supplied to me by the Mathemanian Tourist Bureau is an air-mileage table for the five direct flights which *last* year linked the country's four airports:

	Addon	Taykov
Addon	—	90
Mynis	150	120
Pluz	250	160

Across the bottom is a rubber stamp proclaiming: 'An additional service has now been introduced linking Mynis and Pluz directly.'

They happen to be the two cities I wish to visit, but how am I to tell how far apart they are?

Lead – page 83
Solution – page 107

11 TOURANIAN TOURS

To discover anything about Tourania, one has to pick one's way through the captions which accompany the admittedly stunning photographs with which their government's tourist brochure is studded. Here are three examples:

CITAE, the ancient capital, lies due north of Dockery, the island's main port, and is conveniently equidistant from both Dockery and the new airport at Ayre.

EVERUP, the well-known mountain resort, is only a few miles north-east of Ayre, and lies just twenty miles south-west of the capital. It is easily accessible by a road running south-east to Dockery.

BEACHEY, the famous swimming resort, lying due south of the capital, has recently been linked to Ayre by a new straight road running due east and west across the island.

So what? When I land at the airport, I intend going straight to my seaside hotel at Beachey, and I want to know how far that journey is.

Lead – page 83
Solution – page 107

12 HEARE AND THARE

One headache in these days of supersonic air-travel is caused by jet-lag due to crossing different time-zones. Another source of headache is trying to work out how long a journey is going to take if departure and arrival schedules show local times only.

It can be even more confusing if one's dog has chewed the corner off the schedule, as happened to me when planning my recent flight from Heare to Thare. The time-table now looks like this:

	Normal Service	Supersonic Service
Depart Heare	08.20	09.50
Arrive Thare	17.10	15.40
Depart Thare	20.50	22.50
Arrive Heare	23.40	.....

(The dots represent the dog's tooth marks.)

Assuming that each service (normal and supersonic) each maintains its own regular speed throughout all flights, what time could one expect to arrive back Heare from Thare on the Supersonic Service?

Lead – page 83
Solution – page 107

13 WORLD TOUR

Fig. 5

By now you are probably footsore, jet-lagged, and weary, so why not put up your feet and rest? You can still see a great deal of the world without venturing from your arm-chair.

The maze of twelve letters in Fig. 5 conceals the names of various countries and islands around the world which can be 'seen' by the simple expedient of moving from letter to letter along the lines.

Thus by starting at A, travelling north-west to D, thence east to I, one can 'see' *Adi* (a place I had never heard of until I found it in an atlas while trying to make up this puzzle). But some places are rather better known – *Tahiti* for instance. (No, I have just seen that you can't do that because, although you are allowed to double back for the I-T-I part, there is no direct link between T and A to start with.) Got the idea?

How many countries and islands can you find?

**Lead – page 84
Solution – page 108**

IT'S ON THE CARDS

A pack of problems that call for some degree of mental shuffling, and a trick or two.

14 RED OR BLACK

A deck of 52 playing cards is cut into three separate piles.

In the first pile there are three times as many Blacks as Reds.

In the second pile there are three times as many Reds as Blacks.

In the third pile there are twice as many Blacks as Reds.

How many cards of each colour are there in each of the three piles?

Lead – page 84
Solution – page 108

15 THINK OF TWO CARDS

Prepare a deck of twenty cards: the Ace to Ten of Spades, and the Ace to Ten of Hearts.

Ask someone to select secretly one Spade and one Heart from the deck, explaining to him that an Ace counts as 1.

Now instruct him to:

    Multiply the value of the Spade by 6.
    Add on the value of the Heart.
    Double the result.
    Subtract the value of the Spade.
    Subtract the value of the Heart.

Point out to him that, since he has now subtracted everything he began with, he can hardly be giving away any secret by telling you his final answer.

Nevertheless, when he tells you his answer, you proceed to name both the suit and value of the two cards he has chosen.

Can you work out how that detailed information can be derived from the simple number he gives you?

Lead – page 84
Solution – page 108

16 STRAIGHT ANSWER NEEDED

Fig. 6

Ten, Jack, Queen, King, Ace is as desirable a straight as anyone could wish for. Fig. 6 shows them lying on the table in the order in which they were dealt.

The uppermost card (numbered 5 in the diagram) is of higher rank than the card it is touching, but is of lower rank than the bottom card (numbered 1 because it was the first to be dealt).

The Queen was dealt before the Ten, and the Heart was dealt immediately after the Diamond.

As they lie, the Club and the Ace are separated by two cards, and the two Spades in the hand are separated by the Jack.

Can you name the five cards, according to both suit and rank, in the order in which they were dealt?

Lead – page 85
Solution – page 109

17 HANDY QUESTION

'Examine your hand carefully!' That is the first precept of bridge; so I did!

But since I am a permanent victim of knavish tricks and diabolic luck, I was not surprised to find more Jacks than Trumps, and more Deuces than any other rank of card in my hand.

I had just got as far as calculating that I had more Hearts than Aces, more Queens than Hearts, more Spades than Queens, and more Clubs than Spades, when the dealer barked, 'It's your lead, you know.'

Whereupon, being reluctant to part with an Ace or a King, I defiantly hurled down one of the Fives I held.

We lost the trick – and the game – and the rubber.

Can anyone please tell me what was the composition of my hand? I need all thirteen cards identified by rank and suit. And please, please remind me what were Trumps. My partner has threatened to punch my nose in if I can't remember even that.

Lead – page 85
Solution – page 109

18 WIN OR LOSE

There is nothing like poker when it comes to taking money out of one pocket and putting it into another.

At the start of the game, Jim Straight and Bob Fullhouse between them had four times as much money as Tony Flush, while Tony and Bob together had three times as much as Jim.

At the end of the evening, Jim and Bob between them had three times as much as Tony, while Tony and Bob together had twice as much as Jim.

Bob finished £2 down on the evening.

How much money did each of the three finish with in his pocket?

I don't advise mathematicians to waste time setting up simultaneous equations and all that sort of thing; there is absolutely no need to resort to algebra.

Lead – page 85
Solution – page 110

19 PICK-A-PACK

Fig. 7

The game of PICK-A-PACK is played with sixteen cards – the A,2,3,4 of each suit.

After shuffling, the top six are discarded and the remaining ten are scattered, face up, on the table.

The two players now take turns to claim from the table one or more cards – provided that, if more than one is claimed, the claimed cards must be either all of the same suit or all of the same rank.

The winner is the player whose last claim clears the table.

Assume it is your turn to claim when the play has reached the position shown in Fig. 7.

If you were to claim the three Aces, your opponent could claim the two Diamonds, and you would then inevitably lose. Is it possible for you to make absolutely certain of winning (irrespective of how well your opponent may play) by claiming just two cards – or even just one card?

Lead – page 85
Solution – page 110

20 VERY TRICKY QUESTION

In more ways than one, this really is the trickiest question in the whole of this section.

Our bridge opponents had just made their 'No Trump' contract, and the thirteen tricks lay on the table, ready for the inevitable post mortem.

No trick contained more than one card of the same rank.

No trick contained more than two 'picture cards' (A,K,Q,J).

The 1st trick's highest card was equal in rank to the second-lowest card in the 2nd trick.

The 2nd trick's lowest was higher than the second-lowest in either the 4th or the 12th.

The 3rd trick's lowest equalled in rank the highest in the 2nd.

The 4th trick's second-highest equalled the highest in the 12th.

The 5th trick contained no picture cards.

The 6th trick's highest was lower than the lowest in the 7th.

The 7th trick's highest equalled the second-lowest in the 8th.

The 8th, 10th, and 11th tricks were identical in the ranks of the cards that made up each of them.

The 9th trick's second-lowest equalled the second-lowest in the 5th.

The 12th trick's second-highest equalled in rank the highest in the 13th.

All four of the *Fives* fell in successive tricks.

The very last card to be played was a *Six*.

What ranks of cards made up each of the thirteen tricks? Don't panic – there are only 52 cards for you to identify.

Lead – page 87
Solution – page 110

WHO'S WHO

*Questions of identity for
those who consider personal relationships
more important than material possessions.*

21 WHAT'S IN A NAME?

Father had promised Judy a birthday party.

'And how many guests do you propose inviting?' he asked.

'Forty,' she answered.

'That's rather more than I had been counting on. You couldn't pare it down a bit, could you?'

'I could try,' said Judy.

And in due course she produced her revised list: *Bertram and Yvonne, Stan, Ned, Ian, Alfred, Angus, Amy, Rachel, Enid, Adam, Abel, Sadie, Veronica, Roland.*

The list was duly approved.

So all forty were invited!

The list may appear to contain only fifteen names, but Judy had contrived to work in twenty-five others 'between the lines' so to speak (or perhaps we should say 'between the words').

Herbert for instance (usually called Bert) was conveniently concealed in the name of the very first guest, BERTram.

How many concealed names (being other than mere abbreviations of others on the list) can you discover? For all I know there may be some I haven't thought of, so if you discover more than twenty-five extras, you can go to the party yourself.

Lead – page 88
Solution – page 111

22 FRENCH AND ENGLISH

ROOM 1 – *John and Heloise Smith*
ROOM 2 – *Pierre and Thelma Dubois*
ROOM 3 – *Armand and Françoise Legrand*

That is what the hotel register said.

But not one of the couples had been put into the right room – of that, the hotel porter was quite sure.

'We must sort it out,' said the manager. 'Phone the rooms. Ask them what time they want breakfast. But for goodness' sake, don't ask them their names – we don't want to look utter fools.

'Clearly Smith is an Englishman with a French wife, Dubois is a Frenchman with an English wife, and the Legrands are both French. You should be able to distinguish whether the voice is English or French; and if, after thirty years of service in a discreet hotel, you can't tell whether a voice in a room is male or a female, you are too dim-witted for the job.'

What is the smallest number of phone calls the porter must make in order to discover which couple is in which room?

Lead – page 88
Solution – page 112

23 TABULATION

'Yes, this is the famous round table where we meet,' explained Mrs Georgina Aintwright to the television producer who was planning a documentary on her Hampstead Committee for Discrimination Against Male Propagation (Tittlebats Branch).

'I take the chair of course, and the other five members sit around the table. Mrs Cockshure sits opposite me, and Katherine sits on Mrs Banmen's right. Then there is Mrs Darfter who sits on Norma's left, with Helen sitting between Patricia and Mrs Ennycawse.'

'I thought there was a Mrs Fewtile on the committee,' said the producer.

'Of course there is,' declared the doughty chairperson. 'We couldn't possibly do without her. She sits next to Mrs Banmen – no, that's wrong – she doesn't sit next to Mrs Banmen – she sits next to Janet.

'Now I really have mentioned everyone, and you must excuse me. I am a very busy person – I also run the Campaign for Sending Nuts to the Needy, you know.'

All of which left the producer still trying to work out, not only the full names of the committee, but also just where each sat in relation to the chair – details he must know if he is to site his cameras properly in order to give full coverage to those who are more equal than the others.

Can you supply him with these all-important details?

Lead – page 88
Solution – page 112

24 SMOKO

The conductor and three of the instrumentalists who had just taken part in a major Beethoven work were cooling off in the dressing-room.

The violinist asked Raymond for a light for his cigarette.

'Filthy habit! I have never smoked in my life,' said the cellist.

'I gave up smoking last week,' said Raymond.

'I used to smoke cigarettes once, but now I stick to a pipe,' said Simon.

'Let's not argue about it,' said Quentin, nudging Thomas in the ribs. 'After being on my feet right from the beginning of the performance to the end, what I really feel like is a beer.'

So the pianist, now puffing away happily, led the way to the bar.

What was the opus number of the Beethoven work they had just been performing?

Lead – page 88
Solution – page 112

25 SORTING OUT THE MESS

The five officers in the mess were discussing their Saturday's winnings at the races. Between them:

Simpson and the Engineer had won £15.

Powell and Tait had won £14.

Tait and the Lieutenant had won £12.

Rogers and the Major had won £10.

The Brigadier and the Cavalryman had won £9.

The Artilleryman and the Infantryman had won £7.

The Colonel and the Signaller had won £6.

Quist and the Infantryman had won £4.

The Captain together with the only beer drinker in the party had won £8.

The two whisky drinkers had won £11.

The gin drinker had won more than the brandy drinker.

What was the Rank, Name, Corps, winnings, and drink of each of the five officers?

Lead – page 89
Solution – page 113

26 GOOSEBERRY FOOL

I must, in all fairness, point out that this problem was originally published on 1st April 1978. Many solvers must have let their eyes wander to the dateline at the top of the page, and started to wonder whether or not I was having them on. But I do assure you that the answer is not an April gooseberry.

There is a real answer to the question. Only if you think the problem is logically absurd, should you think of yourself as an April Fool. So here is the puzzle as it appeared.

George and Evelyn never actually met, but they did go on writing almost to the ends of their lives. It has been *said* that Evelyn fell in love with George quite early in life, but I am sure she never even heard of him, despite his fame.

They were both well known to the public, though when it came to making their names, George was certainly the more unconventional – but then he was the more belligerent-sounding of the two.

1928 was the year of his remarkable decline and fall, and Evelyn died in 1966 at the age of sixty-three. A few months after her marriage in 1880, George (by then Cross!) died at the age of sixty-one, but Evelyn did not know this at the time. She had already achieved fame by the middle of March in 1871.

What were George's and Evelyn's surnames?

Lead – page 90
Solution – page 113

FUN AND GAMES

*Games of patience or impatience,
for one or two players.*

27 CATCH PATIENCE

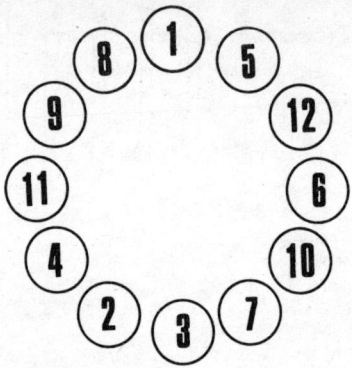

Fig. 8

On twelve discs of card, write the numbers 1–12, and arrange them in a circle in the order shown in Fig. 8, i.e. 1, 5, 12, 6, 10, 7, 3, 2, 4, 11, 9, 8.

Your aim should now be to 'catch' all discs by turning them over like this.

First, turn down any disc you choose. Then start counting from the disc you have turned down, touching each disc as you go. When you touch a disc bearing the number you are just saying, that is a 'catch', so you must turn it down.

For example, if you decide to start by turning down Disc 7 and counting clockwise, you would say 'One' as you touch Disc 3, and 'Two' as you touch Disc 2. This is a catch, so you turn down Disc 2, and set off again – anticlockwise perhaps, in which case you will catch Disc 8 before setting off again. (Notice that Disc 7 was touched even though turned down.)

Can you discover the only sequence of catches that will catch all 12 discs?

Lead – page 90
Solution – page 113

28 ODDS AND EVENS

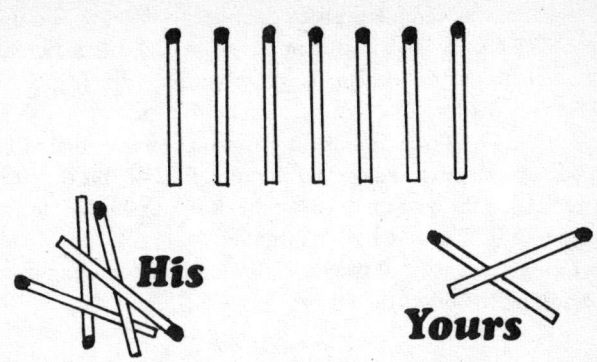

Fig. 9

This is a game for two players. Thirteen matches (or any odd number of matches for that matter) are laid out in a row. The players then take turns to claim either 1, 2 or 3 each time. On each turn, a player *must* of course take at least one.

The winner is the player who, when the table is cleared, finishes with an ODD number of matches. (Whether he has more or less than his opponent is of no consequence.)

Imagine the game has reached the stage shown in Fig. 9, and it is your turn to claim. How many matches (1, 2 or 3) should you claim to make absolutely certain that you can eventually beat your opponent however well he plays?

Lead – page 90
Solution – page 114

29 GRAND CHAIN

The game was complete: all 28 dominoes were now laid out in *Grand Chain*, i.e. with the numbers on each tile matching the touching numbers on its neighbouring tiles (e.g. 2–3, 3–1, 1–5, 5–0).

On the first three tiles of the chain there was a total of 25 spots, on the next four 18, on the next four 28, on the next three 21, on the next three 18, on the next four 43, and on the next four 9. The rest I have forgotten.

Can you identify all twenty-eight dominoes, reading from beginning to end of the chain?

Lead – page 90
Solution – page 114

30 ALL BUTTONED UP

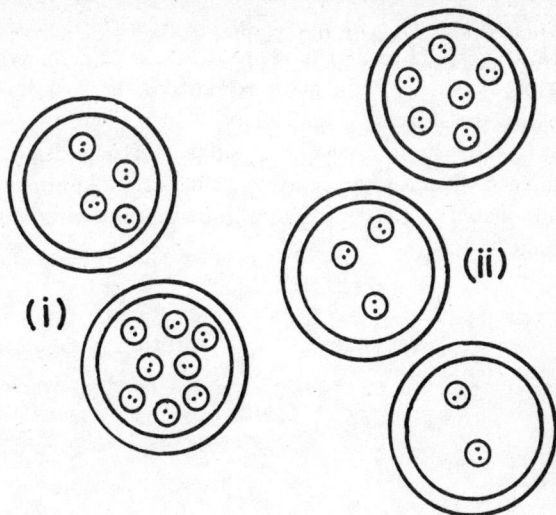

Fig. 10

The game of Wythoff was a popular pastime in the early years of this century.

A handful of buttons is divided unequally between two saucers, and two players take it in turns to draw buttons according to these rules:

On each turn a player must claim EITHER (a) ANY number of buttons from ONE saucer, OR (b) An EQUAL number of buttons from BOTH saucers.

The winner is the player who makes the final draw that clears the saucers.

If a game had reached the stage shown in Fig. 10 (i), and it were your turn to claim, how many buttons should you claim to make absolutely certain of winning?

Try that original version first. Then try a somewhat more tricky proposition. Instead of two saucers, let us have three.

The rules for drawing will be the same except that this time a player must, at each turn, claim:

EITHER (a) *ANY number of buttons from ONE saucer*,
OR (b) *an EQUAL number from two saucers or three saucers.*

Again, the winner is the player who makes the final draw that clears the saucers.

Fig. 10 (ii) shows a possible position that such a game could reach when it is your turn to claim. Again, how many buttons should you claim in order to make quite certain of winning?

**Lead – page 91
Solution – page 115**

31 DIAMOND CHAIN

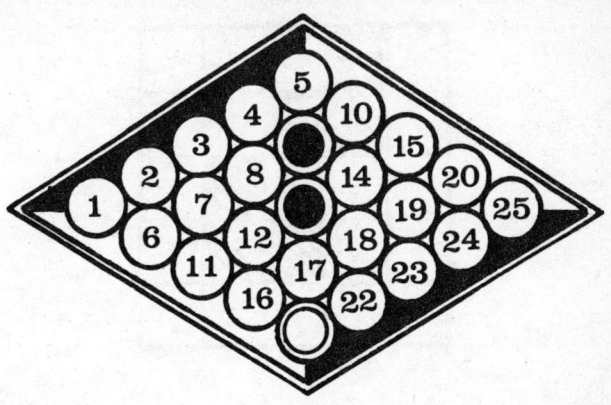

Fig. 11

In this game, one player has a supply of white counters, the other a supply of black, and it is played on a diamond-shaped board as shown in Fig. 11. The two players take turns to place their counters on vacant circles of the board.

White's object is to build a continuous chain between the two white borders, while Black tries to build a chain between the two black borders. For instance, in our example, further white counters on 22, 23, 19, 15 would complete a white chain, while further black counters on 4, 17, 22 would complete a black chain.

You are White, and it is your turn to play. On which circle should you place your next counter if you want to make quite certain of winning, however well your opponent might play?

Lead – page 91
Solution – page 115

32 CHECKEL

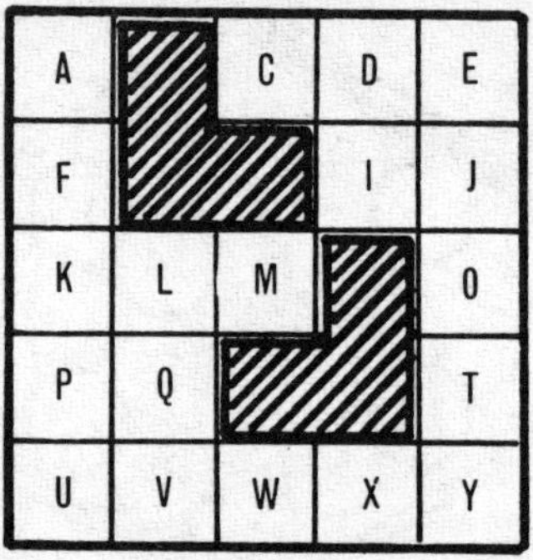

Fig. 12

The game of Checkel is easily constructed. All one has to do is to cut out a number of L-shaped pieces of cardboard each of which will cover three squares of a chessboard.

Two players take it in turns to place one piece at a time in such a way that it covers three squares of the board. The player who finds himself unable to place a piece loses.

The game can be played on an ordinary 8 × 8 chessboard, but to illustrate the game Fig. 12 shows two pieces already placed on a 5 × 5 board.

Of the eighteen positions from which the next player may choose, there are only two that will guarantee a win against even the most experienced player.

What are those two absolutely safe positions?

Lead – page 92
Solution – page 115

33 BREAK'N THE BANK

On the front of six plain white cards, write boldly the six letters, B,A,N,K,E,R – one letter to each card.

Turn the cards over, and on their backs write the names of the six currencies which appear below the letters indicated here:

B	*A*	*N*
MARK	DRACHMA	SCHILLING

K	*E*	*R*
CENTAVOS	DOLLAR	POUND

Ask someone to think of one of these currencies (and make sure that he can spell it properly). He must not of course tell you which one it is that he has chosen. Now turn the cards over so that they again read BANKER.

Tell your victim to spell his chosen word, naming (silently to himself) just one letter of his word each time you tap a card with the end of your pencil.

When he reaches the last letter of his word, he is to say 'Change Please!'

Whereupon, you turn over the card on which your pencil is at that moment resting – and behold, it is the very card which bears the name of the currency he was thinking of.

The puzzle here is for you to fathom out the secret of a tapping system which will enable you to 'break the bank' in this way.

See if you can discover such a system for yourself.

Lead – page 92
Solution – page 116

ON THE SQUARE

*Puzzles which call for the manipulation of
numbers, all of which have something
to do with squareness.*

34 TAKE YOUR PICK

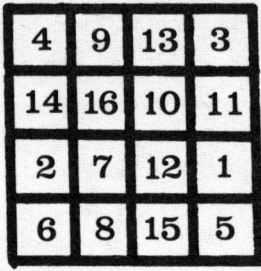

Fig. 13

From the sixteen numbers shown in Fig. 13, choose those four which add up to the greatest total.

'That's easy!' I can hear you say. 'I'll just pick the four biggest – 16, 15, 14, 13. They add up to 58. No other selection can possibly beat that.'

But wait a minute: I haven't finished. No two of the chosen numbers must appear in either the same row or the same column. This means that if you choose 16, you can't also have 14, because those numbers are both in the same *row*. Similarly, should you opt for 15, you can't also have 13, because those two are in the same *column*.

That makes things somewhat more tricky if you are to achieve the biggest possible total.

Lead – page 92
Solution – page 116

35 SQUARE PAIRS

Can you take the first sixteen integers (1–16) and pair them off into eight pairs so that the sum of each pair is a square number. Each pair need not of course sum to the *same* square number (in fact that would be an impossible task). For instance, 1 and 3 would make the square number 4, while 5 and 11 would make 16, but if you choose those pairs you might well finish up with some awkward numbers left over with no suitable partner.

Having done that, try your luck at pairing off into square totals the first 18 integers (1–18).

Can you also similarly pair off into square totals the first 20 integers (1–20)?

Lead – page 92
Solution – page 116

36 EACH WAY ACCUMULATOR

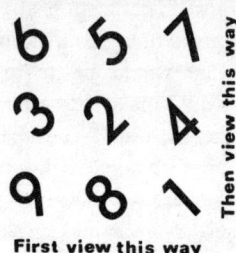

Fig. 14

A very old puzzle calls for the arrangement of the nine digits 1–9 into three lines so that they form a proper addition sum. There are many solutions. One of them is shown in Fig. 14. As you will see, 657 + 324 = 981, and each of the nine digits appears only once.

What does not seem to have been investigated is the possibility of rotating such patterns. If you rotate Fig. 14 clockwise through 90° you will see a somewhat different sum: 936 + 825 = 147. Unfortunately 936 plus 825 does not equal 147.

This raises the question: is it possible to arrange the nine digits 1–9 into three lines of three digits each (similar to the diagram) so that they form a correct addition sum when read normally, and remain a correct addition sum when rotated clockwise as described above?

Lead – page 93
Solution – page 117

37 CROSSNUMBER

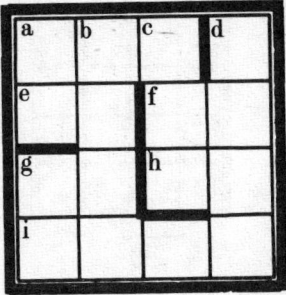

Fig. 15

A crossnumber puzzle is similar to a crossword puzzle, except that instead of the clues and answers being words, they are numbers.

Sometimes a crossword can be completed even if some of the clues are missing. If the pattern is very 'open', all the letters of (say) a Down word might be supplied by the letters of Across words which have already been filled in. So it is with crossnumbers.

The pattern in Fig. 15 requires eleven numbers to be filled in (six Across numbers, and five Downs), but only five clues are needed to identify every digit in the pattern. To do this, you will have to compare some clues with others, even to get a start.

CLUES

1. a $ac = (a\ dn)^2$
2. h $ac = (f\ ac) \times 2$
3. i $ac = (c\ dn) + (g\ dn)$
4. b $dn = (g\ ac)^2$
5. d $dn = (a\ ac) \times (e\ ac)$

Lead – page 94
Solution – page 118

38 PSC SQUARE

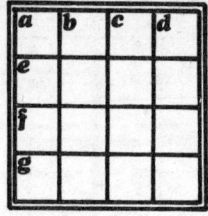

Fig. 16

In this crossnumber, instead of being given detailed clues to the answers, you are merely told what sort of number has to be inserted in the spaces provided.

As you can see from Fig. 16 the eight numbers sought are all four-digit numbers. Each of them is either a prime (i.e. not divisible by any other number), a square (i.e. the result of multiplying some number by itself), or a cube (the result of multiplying some number by itself twice).

The completed pattern must show each of the nine digits 1–9 *at least once*.

There is only one possible solution.

CLUES

	ACROSS		DOWN
a	A cube	a	A prime
e	A square	b	A square
f	A square	c	A square
g	A prime	d	A cube

Lead – page 94
Solution – page 118

39 DIGI-TALLY

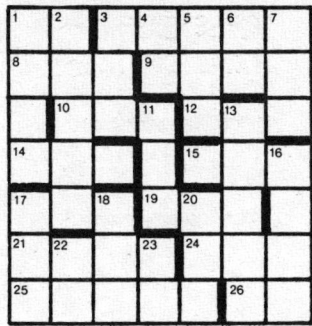

Fig. 17

This crossnumber is based on an entirely different principle. Each clue consists of a bracketed number. That number represents the *sum of the digits* contained in each of the twenty-nine numbers (fourteen of them Across and fifteen of them Down) in the pattern of Fig. 17.

No digit occurs more than once in any number.

And of course no number can start with a zero.

CLUES

ACROSS

1	(5)	15	(17)
3	(19)	17	(17)
8	(21)	19	(4)
9	(7)	21	(6)
10	(10)	24	(13)
12	(14)	25	(25)
14	(9)	26	(14)

DOWN

1	(28)	11	(12)
2	(10)	13	(24)
3	(17)	16	(8)
4	(5)	17	(8)
5	(6)	18	(11)
6	(12)	20	(17)
7	(15)	22	(6)
		23	(11)

Can you complete the pattern?

Lead – page 95
Solution – page 119

MENTAL BLOCKAGE

*Questions concerned with the shaping of blocks,
but which are
quite unsuitable for blockheads.*

40 EXHIBITIONISM

Fig. 18

Despite continuing protests from taxpayers, the Taint Gallery persists in its policy of paying enormous sums of money for weird exhibits.

Fig. 18 shows their latest acquisition – or at least what is left of it. Originally it consisted of a pile of solid, rectangular blocks of bully beef, each measuring 12 × 6 × 6 decimetres.

As you can see, some are now missing from one corner; purloined either by hungry down-and-outs or by public-spirited vandals.

How many blocks remain in the now quite 'high' pile?

**Lead – page 95
Solution – page 121**

41 CUT IT OUT

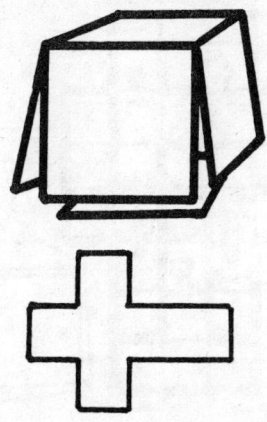

Fig. 19

If seven of the twelve edges of a hollow cube were to be cut after the fashion shown in Fig. 19, and the faces of the cube then opened out, one would finish with a cross-shape as shown at the bottom of the diagram.

A different choice of edges when making one's cuts could produce other shapes.

How many distinguishably different, flat, continuous shapes is it possible to contrive by cutting the edges of such a hollow cube?

(Two shapes may be considered distinguishable if, when rotated or turned over, they still cannot be made to match.)

Lead – page 96
Solution – page 121

42 MAKE IT UP

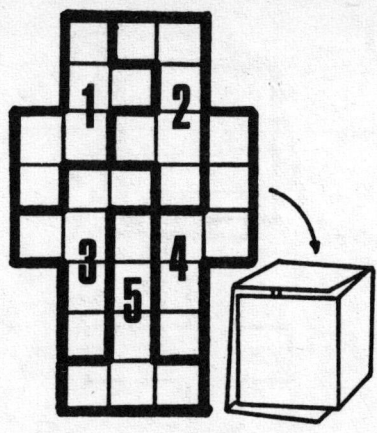

Fig. 20

The previous problem dealt with cutting up ready-made cubes. What about the problem of constructing them in the first place?

I want to cut out an oddly-shaped 30-square-inch piece of cardboard in such a way that the resulting pieces can be folded to make five dice. My first attempt (shown in Fig. 20) was not all that bad; the pieces numbered 1,3,45 can all be folded to form dice, but the piece numbered 2 cannot.

Can you divide the cardboard along the lines into five pieces, all five of which can be folded to form dice?

Lead – page 96
Solution – page 122

43 IN BLACK AND WHITE

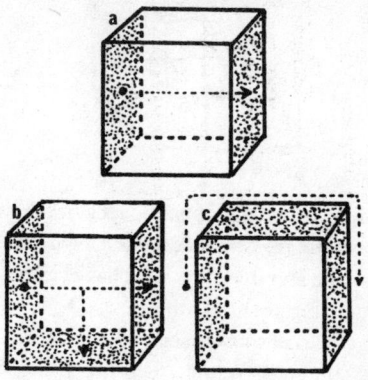

Fig. 21

In how many distinguishably different ways is it possible to colour the faces of a cube black and white? The question is somewhat more tricky than one might expect.

For instance, we could colour two opposite faces black (as in *a* of Fig. 21). That is one way of colouring.

We could then do the same again but blacken in one of the intervening faces (as shown in *b*) to produce a second method of colouring.

Another approach would be to colour three consecutive faces black (as in *c*).

But beware! Were *c* to be inverted, it would look exactly like *b*, so these two methods of colouring are not really different.

How many totally different ways are there of colouring cubes black and white so that they can always be distinguished?

Lead – page 97
Solution – page 122

44 RUBIKITIS

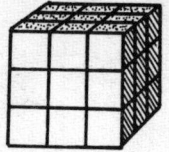

Fig. 22

The problem of colouring cubes is very much more complex when it comes to making every side of a cube a different colour, as anyone should know if he has ever played with one of those infuriating Rubik Cubes.

For anyone who has never seen a Rubik Cube, its sides are traditionally coloured White, Yellow, Blue, Green, Red, Orange, and the object of the exercise is to twist the little blocks until each face shows just one colour, as in Fig. 22. When one starts, the little blocks are all jumbled up, but at least that complication does not concern us here.

However, it did concern me; after two bottles of aspirin I finally managed to get one face of my cube right, and proudly sallied forth to demonstrate my genius to a friend who also has one of the fiendish devices. You see, I had worked out what experts call an algorithm – a sure-fire sequence of moves.

Alas, on my friend's cube, my lovely algorithm didn't work. Eventually I discovered why. In its pristine (i.e. unmucked-up) condition, my cube has Yellow on the side opposite White. His cube was originally designed so that it has Blue opposite White.

All of which leaves me wondering just how many distinguishably different ways there are for a manufacturer to arrange his six colours on the sides of a pristine Rubik Cube.

Lead – page 97
Solution – page 123

45 SYMBOLIOSIS

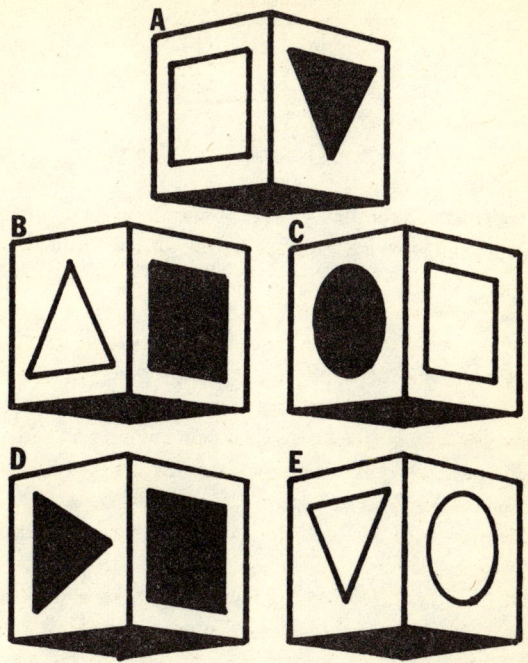

Fig. 23

Fig. 23 shows five different views of *the same* wooden block, the sides of which carry six different symbols.

What symbol would you see on the *top* side of each of the five illustrations if you could peer over the top edge?

Lead – page 98
Solution – page 123

QUID EST VERITAS?

*Well might Pilate have asked that question, for
who indeed knows what is truth? The truth of the matter
is that it is not always easy to decide the truth of the matter,
and establishing the truth of the matter is what
this final section is all about.*

46 ROYAL DILEMMA

'I have determined that you shall marry either Baron Pompuss or Count Grewsumm,' said the king to his daughter. 'It is my will that, if you prove to be a truthful princess, you shall marry Pompuss and him only, for he deserves a loyal and honest wife. If anything you say should prove to be false, you shall marry Grewsumm and none other, for he alone knows how to tame a shrewish woman. I will allow you to make just one statement before I finally decide which of these two fine noblemen shall be your husband, lord, and master.'

The king glared.

Pompuss preened himself.

Grewsumm leered.

The poor princess blanched with fear, and trembled.

What simple statement could she possibly make that would enable her to escape the loathsome prospect of being wed to either of the two repugnant nobles who had set their ambitious hearts on a royal alliance?

Lead – page 98
Solution – page 124

47 AGE OLD QUESTION

Women may not be quite so reluctant to discuss their ages as they were in grandmother's day, but when it comes to that delicate question their veracity is not always to be relied upon with complete confidence. Truthful or not, they can certainly wax indignant about insinuations.

'Whatever you say, I am certainly not over forty,' declared Margaret.

'You are at least five years older than I am,' snapped back Penelope, 'and I am thirty-eight.'

'You?' sneered Margaret. 'You are at least thirty-nine.'

One could be charitable, and assume that they really had forgotten their ages, but I happen to know for a fact that not one of the statements they made is correct.

How old are they really?

Lead – page 98
Solution – page 124

48 AMPHIBOLIA

Traditionally the Amphibolian caste system was strict, the entire population being classified as either Fellahs or Phibbas.

By hallowed tradition, Fellahs invariably spoke the truth, where Phibbas invariably lied (even as to their own caste). Because native Amphibolians could recognise caste instinctively, no confusion arose locally, but things could prove difficult for a visitor, as I discovered on my first visit to the country some twenty-odd years ago.

Are you mostly Fellahs or Phibbas?' I asked the five Amphibolians with whom I was drinking in the bar of my hotel.

'More of us are Fellahs than Phibbas,' replied the first.

The second shook his head. 'More of us are Phibbas than Fellahs,' he said.

I looked perplexed.

'Only two of us are Fellahs,' explained the third.

'Only three of us are Phibbas,' added the fourth.

'And what of you?' I enquired of the fifth man.

'I am a Fellah,' he replied.

Which of the five men were really Fellahs, and which were Phibbas?

Lead – page 99
Solution – page 124

49 BUT WHICH OF US?

After the episode described in the previous puzzle, it was not until last year that again I visited Amphibolia. The two traditional classes of Fellahs and Phibbas still exist, but since my previous visit a revolution had taken place, resulting in the emergence of a third class of citizen. These middle-class egalitarians (known as Jokers) eschew the ancient conventions of caste, and religiously alternate truth with falsehood. If one statement made by a Joker happens to be the truth, his next will be a lie, and vice versa. The trouble is, that when one encounters a Joker, it is difficult to know which truth/falsehood leg he happens to be on at the time he first opens his mouth.

I was in Amphibolia for the purpose of writing an article on the social changes that have taken place since the revolution, and was met at the airport by three men sent to greet me on behalf of the Ministry of Disinformation. I tried the same opening gambit as on my former visit.

'How many of you are Jokers?' I asked.

'Just one of us,' volunteered Xzymith.

'At least one of us,' added Yxzmith unhelpfully.

'More than one of us,' said Zxymith.

'Are you yourself a Joker?' I asked Xzymith.

'No,' he replied.

'He is,' said Yxzmith.

'He isn't,' declared Zxymith.

Which, not surprisingly, left me wondering which caste each of the men really belonged to. Can you throw any light on the matter?

Lead – page 99
Solution – page 125

50 TELL ME TRULY

My most distressing Amphibolian experience came on the same evening as the conversation described in the previous puzzle.

At my table in the hotel dining room were three Amphibolians staying as fellow guests: two men, and one (very attractive) girl whom, to be honest, I had been trying to chat up earlier but, being ignorant of her caste, was still in the dark as to whether my affections were truly reciprocated.

In a desperate effort to settle my doubts, I tried to establish what social conventions each of my table companions subscribed to.

'I take it that your friend Ambrose here is a Fellah,' I said to the beautiful, seductive, elusive Caroline.

'Ambrose is certainly no Fellah,' she laughed. (Such a delightful, tinkling laugh she has.)

'Caroline is a Phibba,' explained Ambrose politely, 'and Bertram here is a Joker.'

'Ambrose is a Phibba,' corrected Bertram. 'Caroline is a Joker.'

Again Caroline smiled, then suddenly laying her hand on my arm, she whispered confidentially, 'Whatever they say, I really am in love with you.'

My head swirled!

After that I couldn't care less what the men were. I lay awake all night wondering whether the sweet young thing really did respond to my feelings or not. I still dream of her. Please, please, can someone tell me whether she loves me or not?

Lead – page 99
Solution – page 125

THE LEADS

The purpose of this section is to provide the solver with hints or clues that may assist him in solving a problem which proves elusive in its originally stated form. In some cases the Lead may also indicate what is *not* the answer, so that if perchance, on consulting the Lead, the solver should discover that he has arrived at a wrong answer, he can go back and check his reasoning before reading the full solution and answer.

To compress the length of the reasoning, certain mathematical and logical conventions are used, and these are listed below.

+	'plus' or 'in addition to'
−	'minus' or 'without'
×	'multiplied by'
/	'divided by'
=	'equals' or 'is the same as' (e.g. A = 23 may, according to the context, mean 'Albert is 23 years of age'.)
< >	'does not equal' or 'is not the same as'
>	'is greater than' or 'takes precedence over' (e.g. A > B may, according to the context, mean 'Andy is taller than Bill'.)
<	'is less than' or 'takes precedence below'
= >	'equals or is greater than'
= <	'equals or is less than'
ABC	'Those three items in that fixed order'
A,B,C	'Those three items respectively'
A'B'C	'Those three items, but not necessarily in that order'
A;B;C	'A choice of those three items'
?	'Some digit or letter or item not yet identified'

The same conventions are used in the Solutions.

1 *A Little Choosey*

Use B for caBBage, R for caRRots, L for cauLifLower. Now of the 4 who eat both cabbage and carrots, 1 is obviously our omnivorous B+R+L, leaving 3 who eat *only* B+R (and therefore do not eat L). Similarly there are 2 eating only R+L (and not B), and 1 eating B+L (and not R). These data can be entered on a table thus:

WILL EAT	NUMBER	NUMBER WHO WILL NOT EAT		
		B	R	L
B+R+L	1	0	0	0
B+R	3	0	0	3
R+L	2	2	0	0
B+L	1	0	1	0
B				
R				
L				
Nothing				
TOTALS	?	5	5	5

Since (according to the second paragraph of the puzzle) each of the last three columns must sum to 5, a further 3 must be inserted somewhere into the missing entries in column B, 4 into column R, and 2 into column L. For instance, of the 4 who (in the story) will not eat R, there must be 1 who WILL eat B (and only B, since those eating B+L have already been tabulated). What of the 2 who will not eat L?

2 *Poulterer's Dozen*

Let d = the price per dozen. We then have: change received for a dozen = 50 − d, and for 50p we could buy 50/(d/12), i.e.

600/d. These values are the same, so $50 - d = 600/d$. If you despise trial and error you must now resort to quadratic equations, but I must warn you the price was not 20p per dozen.

3 Mix Up

Tabulate weights thus (with w representing the unknown weight of an egg).

(a) Ingredients	(b) Book	(c) My mix	(d) My proportion must	(e) New mix
F(lour)	9	10	Fall	
O(atmeal)	10	9	Rise	
B(utter)	8	6	Rise	
S(ugar)	6	8	Fall	
E(ggs)	6w	6w	Same	
TOTALS	33 + 6w	33 + 6w	Same	

No ingredient can be reduced in quantity, therefore must be increased or left constant. Clearly if one ingredient can be left constant, the total final mix will be desirably lower, for if *that* ingredient were increased, *all* others would have to be proportionately larger.

Can E be left constant (irrespective of the standard weight of an egg)? What about leaving F,O,B or S constant?

4 Measure for Measure

This is not a trial and error problem, but a fairly simple exercise in algebra. Let the *capacities* of the jug, flask, and bowl be j,f,8. Then set out the *contents* of the vessels at each step thus:

STEP	JUG (j)	FLASK (f)	BOWL (8)
0 (Start)	j/2	f/2	0
1	0	f/2	j/2

What quantities of milk are in the flask and bowl after the fourth and final step?

5 *Well Done*

There is no mathematical algorithm for solving this sort of problem. It calls for sheer ingenuity. If you can get the time down to 11 minutes, you can certainly say 'well done', or call it a 'rare success'; at least that is what I thought when I tried my hand at it. But my invigilator Peter Mabey showed that even that time can be improved upon, provided no diner insists that each side of a steak must be cooked *uninterruptedly*.

6 *A Piece of Cake*

Fig. 24 shows the three-cut dissection which I suggested in the newspaper when I published my solution.

Mark x,y,z (the mid-points of three sides). Cut from D to x, and from C to y. Then cut from A towards z, but only as far as the line Dx. The pieces can then be fitted together to form two square cakes of the same thickness, the smaller being a quarter of the area (and therefore a quarter of the weight) of the other.

Mathematically minded solvers seemed happy enough with this proposal; practical housewives were furious that I should have suggested such a 'crumby' solution.

'With three cuts, you could make a far more straightforward job,' I was told, 'a method far less likely to fall to pieces when the cake is finally iced.'

They were right, so try to discover a more *practical*

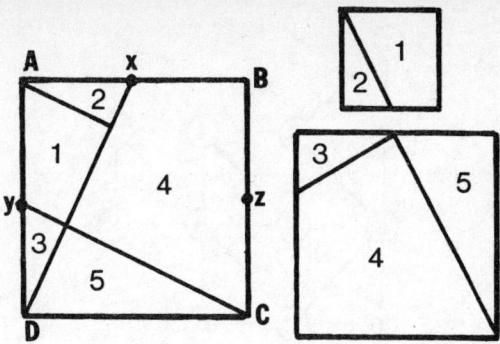

Fig. 24

approach to the problem (and try to keep your suggestion 'on the level' so to speak).

7 *Far Enough*

What proportion of our journey had we covered at the time of the first question, and what proportion at the second? What *proportion* of the journey separated these two points, and by what *distance* were they separated?

8 *Local Colour*

Because it touches Areas 3 and 5, Area 2 on the map cannot be B or R, so must be Y or G. What about Area 4? How does this affect Area 6? Are there any areas towards the east of the map which lead similarly to the firm identification of some other area?

Having established this, now consider Area 1.

9 *Mystery Tour*

Distort the diagram orthogonally as in Fig. 25, and list the tours thus:

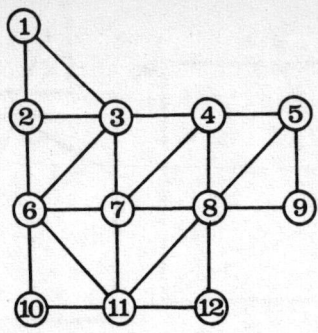

Fig. 25

B – K – J – T – C – A – L – B   (7 links – ODD)

B – L – R – M – S – D – F – B   (7 links – ODD)

B – J – C – A – R – D – B   (6 links – EVEN)

Now a tour with an odd number of links cannot reach home unless it uses one diagonal (or an odd number of diagonals).

Since all 12 towns are mentioned in the two ODD tours, one of these ODD tours (call it X) must include Town 1, and must therefore use the diagonal Link 1–3. With only seven links for each of the ODD tours, this tour X cannot reach 5 or 12. These towns must be covered by the other ODD tour with a diagonal (call this tour Y), but diagonal 5–8 would cut off 9, and diagonal 8–11 would cut off 12. Links 3–6, 6–11 are too distant from 5 and 12 to be included in the same ODD tour, leaving us with only 4–7 as the diagonal for tour Y if 5,9,12 are all to appear in the same tour. This gives the cyclic order of Tour Y as 4–7–11–12–8–9–5–4.

Forget about trying to letter the towns for the time being. Concentrate on deciding what the other seven-link Tour X must comprise.

10 *Air Miles*

What distance is PT + TA? What distance is PA? What can you say about the route PA? In what proportions are TA, MT, MA? Does that ring a school bell?

11 *Touranian Tours*

The geography must be as shown in Fig. 26 (in which the angles have been entered). The only actual distance given is

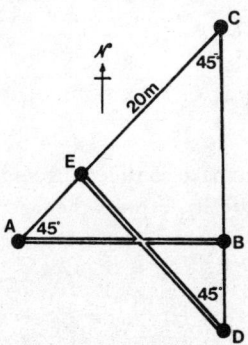

Fig. 26

20 miles from E to C, but certain congruences can be deduced which will enable you to calculate A to B.

12 *Heare and Thare*

What is the *apparent* time saving (due to time-zone variations) between the outward and homeward flights of the Normal Service? What will be the *apparent* time saving for the Supersonic flight?

13 *World Tour*

In this sort of problem all you can do is use your initiative, atlas, and gazetteer. You can count 12 as a pass, but there are probably 30 or more if some very obscure islands are counted.

14 *Red or Black*
Set out your data thus:

Pile	RED	BLACK
1	x	3x
2	3y	y
3	z	2z
TOTALS	26	26

This will help you make considered guesses, but if you want to work mathematically, express the facts this way:

$$\text{RED} \qquad x + 3y + z = 26$$

$$\text{BLACK} \qquad 3x + y + 2z = 26$$

and treat this as though you were solving a simultaneous equation. Subtract twice the RED equation from the BLACK, and you will get $x = 5y - 26$, from which it is clear that y cannot be less than what number? Now try subtracting the BLACK equation from three times the RED. What does this teach you?

15 *Think of Two Cards*
Let S = Spade value, and H = Heart value. Now, step by step, work out what *total* your victim reaches; e.g. Step 1 will give 6S, Step 2 6S + H, and so on. By what number will the first term in the final result divide? If the *whole* answer is divided by that number, what will be the quotient and remainder?

16 *Straight Answer Needed*

Calling the ranks A,K,Q,J,T, the suits s,h,d,c, and the cards by their diagram numbers, set out your data thus:

(i)   1 > 5
(ii) 5 > 4
(iii) T above Q
(iv) h immediately above d
(v)   c??A (either up or down)
(vi) sJs (either up or down)

By (i) 1 < > T, and 5 < > A. What can you deduce from (ii), (vi), (iii), (v)? If J = 2 or 3, what about (iv)?

17 *Handy Question*

Call the Deuces X, the Fives F, and the Trumps t. Then (so far as quantities are concerned) we can say: c > s > Q > h > A, therefore Q = > 3. So since X > any other rank, X = 4 (one of each suit) and Q = exactly 3.

If A = 2, what is the smallest number of suit-named cards we could have? What is the smallest possible number of J? How many K? Decide how many of each rank there must be, then try to decide their suits.

18 *Win or Lose*

What fraction of the trio's total pool did T win? What fraction J? What fraction did B *lose*? How much then was the pool?

19 *Pick-a-Pack*

Fine judgement and a lot of luck are required to win this game in practice but, towards the end, close analysis is possible. If the suits and values are tabulated as in Fig. 27, certain combinations can be proved 'Safe to leave':

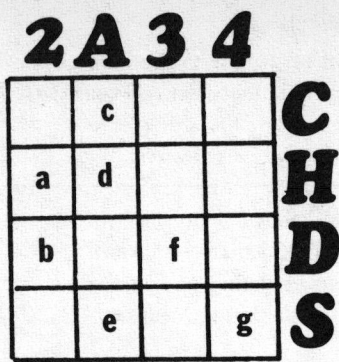

Fig. 27

(P) TWO cards if they are unlinked (i.e. do not share a row or column) like df.

(Q) TWO unlinked pairs, like de,bf; ab,eg; cd,bf; bf,eg.

(R) FOUR unlinked cards, like cafg.

(S) FIVE forming two pairs with one link-pair, plus one odd card, like ad,bf,g, the link-pair being ab.

(T) SIX cards forming two L-shaped triplets with no link between them, like abf,ceg.

P,Q,R are obviously SAFE.

Draw for yourself a diagram of S, showing only the five cards a,d,b,f,g. Then such an array is SAFE, because a(bf)P, which means that if HE takes a, then YOU take (bf) leaving P (which is again SAFE).

Similarly, S is SAFE because, b(ad)P, d(ab)P, f(ab)P, g(ab)P, ad(b)P, bf(a)P, ab(d)P.

Draw a diagram of T, showing a,b,f,c,e,g. Then T is SAFE because a(c)Q, b(e)R, c(a)Q, e(b)R, f(c)Q, g(a)Q, bf(ce)P, eg(bf)P, ab(ce)P, ce(ab)P.

So see how you can reduce the seven cards shown in the problem to one of these SAFE combinations.

20 Very Tricky Question

The reasoning is very complex, and only an outline of it can be given, but it will help if you draw yourself a diagram with 13 rows (representing Tricks 1–13) and 4 columns headed L,NL,SH,H (representing Lowest, Next Lowest, Second Highest, Highest in each trick). Then, starting with paragraph 3, letter the paragraphs (a)–(n).

Then by para (b) T8(SL) = < 10, so by (i) T7(H) = < 10. So, taking (h) into account, we can enter for Tricks 6,7,8:

	L	NL	SH	H
(6)	2;3	3;4	4;5	5;6
(7)	6;7	7;8	8;9	9;10
(8)		9;10		

Now the first trick's highest cannot be less than 5, so by (c) T2(SL) = > 5, so by (a) T2(H) = > 7, so by (e) T3(L) = > 7, so T3 does not contain a 5, and by the table above T7 does not contain a 5.

Assume T6 contains a 5, then by (m) 5s must occur in Tricks 3,4,5,6; 4,5,6,7; 5,6,7,8; 6,7,8,9, but we have just proved that neither T3 nor T7 contains a 5. So, taking (h) into account, we can definitely assign T6 and T7. And since by (i) T8 now contains a 10 as NL, so by (j) do T10 and T11 as well. So we have:

(6)	2	3	4	6
(7)	7	8	9	10
(8)		10		
(10)		10		
(11)		10		

There is a long way to go, but 'Courage brother! Do not stumble!'

21 *What's in a Name?*

This is just a question of knowledge and observation, but don't forget that new names may span old ones, e.g. *Mandy* appears in Bertra*m and Y*vonne.

22 *French and English*

For each room, try to think what you could deduce from hearing a male or female voice, combined with a French or English accent. Can any room yield more information than the others?

23 *Tabulation*

Draw a circular diagram, marking the six chairs clockwise (i) to (vi), assigning (i) to the chairman. List the data thus: (1) gA = chairman; (2) C opp A; (3) k on B's right; (4) D on n's left; (5) h between p and E; (6) F not next to B; (7) F next to j; (8) already identified (this being a convention we shall use later).

Then by 1, (i) = gA. By 2, (iv) = C. By 8, (vi) < > g. By 5, (vi) < > h,p. By 3, (vi) < > k. By 4, (vi) < > n. So (vi) = j. Then by 7, (v) = F. By similar reasoning either a surname or a first name can be identified if the chairs are considered in the following order: (ii), (iii), (iv), (vi), (iii), (ii), (v).

24 *Smoko*

The cellist (who never smoked) cannot be R (who gave it up last week), nor can he be anyone who still smokes or stands up to play. Who then can he be? What of the pianist? What is R? What is Q?

Without an encyclopedia or an encyclopedic knowledge of Beethoven's works, the ultimate question cannot really be answered, but you can claim a 99 per cent mark if you manage to identify which instrument each man plays.

25 *Sorting out the Mess*

Tabulate the men (in descending order of wealth) a, b, c, d, e.

Now since ten *different* totals are given for only five men, no two men can have won the same amount, and each man has been counted four times, so the total money won is 1/4 of the combined total of all monies mentioned (£96) i.e. £24. Since $a+b=15$, and $d+e=4$ then $a+b+d+e=19$, leaving $c=5$. Now since $a+c=14$, $a=9$. Since $c+e=6$, $e=1$; since $a+b=15$, $b=6$; since $d+e=4$, $d=3$.

Then the winnings are $a+b=15$ (S+Eng), $a+c=14$ (T+P), $a+d=12$ (Lt+T), $a+e=10$ (Maj+R), $b+c=11$ (both Whisky), $b+d=9$ (Brig+Cav), $b+e=7$ (Art+Inf), $c+d=8$ (Capt+Beer), $c+e=6$ (Col+Sig), $d+e=4$ (Q+Inf).

Since $a+c=T+P$, and $a+d=T+Lt$, the common name T must be a; so $c=P$ and $d=$Lieut. Since $a+b=S+$Eng, and $a<>S$, $b=S$ and $a=$Eng. So we can now make several entries on a table thus:

	Win	Rank	Name	Corps	Drink
a	9		T	Eng	
b	6		S		W
c	5		P		W
d	3	Lt			
e	1				

Work now to identify (in this order) Brig, Cav, Sig, R, Maj, Q, Inf, Art, Col, Capt, Beer, Gin, Brandy.

26 Gooseberry Fool

Did anyone suggest that Evelyn was a woman? Has there never been a woman called George? Have you no taste for literature? Gibbon was not the only man to write a *Decline and Fall*, and someone did once write about *Middle March*.

27 Catch Patience

Since 12 can be reached only from 12, *it* must be the 'free turn', leading inevitably to 10 (anticlockwise), and 4 (clockwise) as the next two catches. Since 6 can be reached only from 11 (clockwise to avoid the intervening 3) the sequence, 11,6,3, must appear somewhere. Since 10 is already accounted for, 3 can lead only to 5. From 4 one can reach 5 or 11, but we have shown that 5 must be accessed from 3. You should now be able to write down the first seven catches. Try to go on from there.

28 Odds and Evens

If you take 1 match, and he were to take 2, who then would win? What might your opponent do if you were to take 3?

29 Grand Chain

This is a very difficult problem. Index the groups with letters *a–h*, showing the number of tiles (in brackets) followed by the number of dots: $a(3)25$, $b(4)18$, $c(4)28$, $d(3)21$, $e(3)18$, $f(4)43$, $g(4)9$, totalling 25 tiles with 162 dots. Then, since an entire set of dominoes consists of 28 tiles with 168 dots, the last group must be $h(3)6$. It will also help if you tabulate all 28 tiles (6–6, 6–5, 6–4 ... 0–0) and cross them out as you assign them.

Now the dots on the tiles of the *gh* groups total 15. Your table will show that *gh* must be made up of tiles that reveal no dots other than 4,3,3,2,2,1,0. It will also show that the four *f*-group tiles totalling 43 can be only the 6–6, 6–5, 6–4, 5–5.

Now the highest dot appearing in *gh* is a 4, and the lowest in *f*-group is a 4; so two 4s must form the link between the two groups, giving 6–4, 4–0 (with the remaining tiles of *gh* being 3–0, 2–1, 2–0, 1–1, 1–0, 0–0). But the 3 here has no other 3 in that group to link with, so the 3 must be the last dot of the chain. Therefore we can establish *fgh* as 5–5, 5–6, 6–6, 6–4, 4–0, 0–0, 0–2, 2–1, 1–1, 1–0, 0–3.

Now it can be shown that in *any* complete Grand Chain, the very last dot matches the very first, so since the chain ends with 0–3, it must start with 3–?. So off you go from the front end.

30 *All Buttoned Up*

To give an exhaustive analysis of this game would take many pages, but the winning strategy in *all* games is to reduce the buttons to two saucers containing 2,1.

With two saucers (and up to 8 buttons in a saucer) the following combinations are the *only* ones SAFE to *leave*: 7–4, 5–3, 2–1, 0–0. Whatever your opponent claims when confronted by such a situation, you can then either clear the saucers or reduce the numbers he leaves to one of the lower SAFE combinations.

In the case of three saucers (with up to 6,3,2 buttons) the only SAFE combinations to *leave* are: 6–2–2, 3–3–1, 4–1–1 (plus the last three of our old two-saucer SAFES which can now be expressed as 5–3–0, 2–1–0, 0–0–0). Again, after your opponent's move, aim for another lower SAFE combination.

31 *Diamond Chain*

Your aim should be to claim circles which give fork-outlets, e.g. if you claim 19, you have a fork on 15;20 which Black cannot stop in one move.

Assume White plays to Circle 5. Black could reply by claiming 18, giving him a fork on 22;23. White must then claim 4 (otherwise Black could claim it and, on his next move, win on either 22 or 23). With White holding 4 and 5, could Black still contrive a win?

32 *Checkel*

A player must aim for a position which will enable him ultimately to leave two *independent* playable spaces on the board. If you play CDI, your opponent could reply KLQ, leaving himself the two independent corners PUV and TXY, so he will win.

Analyse each possibility, to see if *you* can leave such corners.

33 *Break'n the Bank*

Notice that each currency name has a different number of letters, ranging from four to nine. Write the names down in ascending order of length. What letters appear on the backs of these cards? If, when your opponent says 'Four', you should happen to be tapping B, you would win! Which letters should you be tapping when he says 'Five', 'Six', etc? Surely you can think of some way to make sure that your fourth tap lands on B.

34 *Take Your Pick*

If you start with $16+15$, the other possible pairs ($4+1$ or $2+3$) each total 5, giving a maximum answer of 36. Similarly $16+13$ leads to 36. But $16+12$ leaves pairs ($3+6$ or $4+5$) which bring the grand total to 37 (the best one can do if 16 is included). If 15 is our highest choice, $15+14$ will leave pairs totalling 10, giving a grand total of 39. But it is possible to do even better than this, so keep at it.

35 *Square Pairs*

Consider the first problem (1–16). Since the largest possible total is 16 + 15 = 31, the only square totals we can aim for are 4,9,16,25. Clearly 16 can aim for nothing less than 25, so 16 needs 9. With 9 used up, 7 cannot reach the square total 16, so must aim for 9, so needs 2. With 16,9,7,2 assigned, what will be needed by 14 to make a square total? Now seek pairs for 5,12,8,15,6, in that order.

Apply similar reasoning to the problems of 1–18, and 1–20 (remembering in this last case that, since the maximum choice of 20 + 19 equals 39, we can look for square totals of 4,9,16,25,36). Try your luck!

36 *Each Way Accumulator*

If you despise trial and error, and insist on rigorous proof, be it on your own head. Here goes! Assign letters to the array thus:

a	b	c
d	e	f
g	h	i

The sum of the nine digits is 45. The Digital Remainder of 45 is 0 (the DR being the remainder left after dividing by 9). Since abc + def = ghi, the DR on both sides of that equation must be 0. So the DR of ghi = 0. If $g+h+i=9$, then $g'h'i = 1'2'6$ or $1'3'5$ or $2'3'4$. But since $i = g+h$ or $g+h+1$(carried), neither $1'2'6$ nor $2'3'4$ provides suitable values for $g'h'i$. In the remaining case ($1'3'5$), since $a'd = > 1'2$, then $g = > 3$. So the order of ghi would have to be 315 (when we remember that i, being the starter of the total ifc, must be $= > g+h$). Then $g = 3$, making $a'd = 1'2$, but 1 has already been assigned to h. Therefore if any solution exists, then $g+h+i < > 9$; so to have a DR of 0, $g+h+i$ must be 18 (not 27 or more for no three digits sum to that much).

The only possibilities for g′h′i are therefore: 9′8′1, 9′7′2, 9′6′3, 9′5′4, 8′7′3, 8′6′4, 7′6′5. But since i = g + h or g + h + 1(carried), the last three groups are inapplicable. In the first four i must in each case be 9. Then (remembering that g = >3) the only possibilities for ghi are: 819, 729, 369, 639, 459, 549.

Consider each of these possibilities.

37 *Crossnumber*

From (1) since a?? = a?² the first digit must be 1. The greatest possible total for c(dn) is 999, and for g(dn) is 99. Therefore from (3) i(ac) must be 10??. To reach this four-digit number, c(dn) must be 9??, making a(ac) 1?9 which, by (1) is the square of 1?. So a(ac) must be 169 and, by 1, a(dn) = 13. This makes b(dn) 6??0. Since (4) says this is a square number, it must be 6400. Then from (4) g(ac) = 80. From (5) you can now calculate d(dn). What is the last digit of c(dn), the second-last of c(dn), the second-last of i(ac)?

38 *PSC Square*

Consider the cube in d(dn). Its four digits must be legitimate terminal digits for a cube (any digit), a square (0,1,4,5,6,9), another square, and a prime (1,3,7,9). Of the twelve four-digit cubes which exist only 4913 and 2197 fill the bill. If we take 4913, the only four-digit cube ending in 4 to make up a(ac) is 2744, whereas if we take 2197, the only four-digit cube ending in 2 is 5832. So we have only two possibilities to consider:

	2	7	4	4		5	8	3	2
			p	9				p	1
(x)			q	1	(y)			q	9
				3					7

Consider the array (x) in relation to e(ac) and f(ac). If a square ends with 9 or with 1 its penultimate digit (p or q) must be even, so we can try 0,2,4,6,8 for p, and the same for q. All combinations for 4pq? in c(dn) can be rejected on grounds such as 'no square has the form 400?, 402?, 420?'. The closest we can get is with 422? accompanied by rows of 2744, 5329, 7921, ?653, but if the (still deficient) 8 replaces the ?, 2578 is alas not a prime. Ultimately all combinations for array (x) fail, and we must try array (y) in the same way.

39 *Digi-Tally*

The four different digits of 9ac total only 7, and can therefore be only 0'1'2'4. Since 6dn attains 12, it will need the 4 of 9ac, plus an 8 as its first digit, so *6dn = 84*.

Now 3ac with an 8 but no zero (since all its digits start down-numbers) can have in addition to its 8 only 1'2'3'5 to total 19. Then with the digits available from 3ac and 9ac, we must have *4dn = 32* (not 14 because the 4 of 9ac is used in 6dn, nor 50 because zero cannot start 9ac).

Now 7dn totals 15. Its last digit cannot be > 9, so its first two must total at least 6. 5 and 1 are the only digits still available from 3ac and 9ac which can reach this total, so *7dn = 519*.

12ac must be 1'4,9 or 2'3,9, but 5dn totals 6, and since its first is not > 2, *12ac = 419*, making *3ac = 13285*, and *5dn = 204*.

Continue cross-reasoning in this way, aiming to establish in order: 3dn, 8ac, 1ac, 10ac, 14ac, 17ac, 22dn, 21ac, 23dn, 25ac, 19ac, 20dn, 24ac, 26ac, 13dn, 15ac, 16dn.

40 *Exhibitionism*

Think, not of blocks, but of half-blocks (i.e. 6cm cubes). Calculate how many of these were originally in the pile, and how many are missing.

41 Cut It Out

The best way to tackle this problem is to work back-to-front, i.e. decide how many six-square shapes will fold into a box. This is largely a matter of trial and error, but it may help to know that there are 35 basically different shapes that can be formed by joining six squares together. These range from 'all six in a line' to 'all six cramped together in a 3×2 rectangle' — but neither of these arrangements can be folded into a cube. Indeed, less than a third of the 35 can be so folded. Any that can be so folded can also of course be unfolded from a cube.

42 Make It Up

It follows from the previous problem that the only 11

Fig. 28

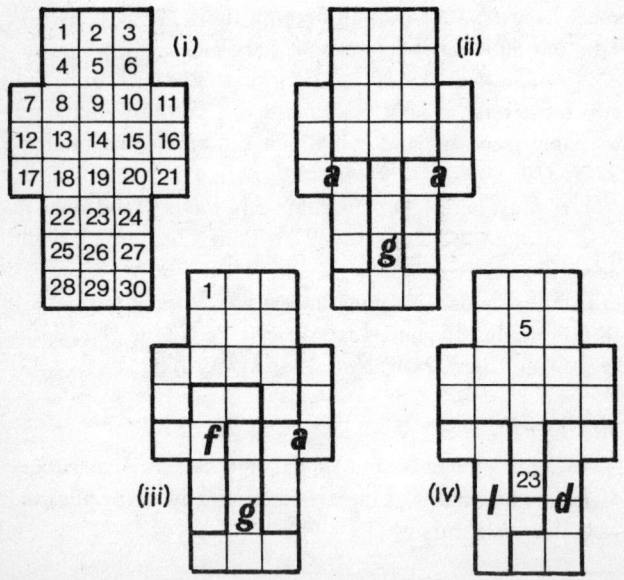

possible shapes to choose from are those shown in Fig. 34, and to these we shall refer by letter. Number the squares of the cross as shown in Fig. 28(i). Now consider corner 28. h;e cannot reach into this corner, and b;c;f;j;k cannot cover it without isolating 30, while a cannot cover it without leaving the impossible four-square area 26,27,29,30. Can g cover 28? Only if it is used upside down. Only a;f;g could then reach 25 or 27 but if we used g it would isolate 17 or 21.

Fig. 28(ii) shows what happens if we use g plus a,a. The remaining rectangle is uncoverable. Fig. 28(iii) shows g plus f,a. Then 1 could be reached only by b;k;j, all of which will leave useless configurations. So g does not account for 28 or (by reflection) 30. To cover these we must use the remaining d and 1, as in Fig. 28(iv). What shapes can now reach 23? What sort of area does each leave uncovered?

43 *In Black and White*

Six faces can obviously be anything from all black to all white, giving seven basic possibilities: (a) 6B, (b) 5B+1W, (c) 4B+2W, (d) 3B+3W, (e) 2B+4W, (f) 1B+5W, (g) 6W. Given these possible ways of sharing out the colours, in how many distinguishable ways can a,b,f,g be arranged? In the cases of c and e, can different arrangements be distinguished by adjacency? What about case d?

44 *Rubikitis*

Any cube can be rolled around until White is on top, so consider only a cube with a White top. Now you have five colours to choose from for the bottom, so top and bottom alone will distinguish at least five cubes. In each case you are left with an 'equator' around the sides (call them p,q,r,s) which can be used for distinguishing cubes which have the same bottom-colour. How many ways can you arrange p,q,r,s?

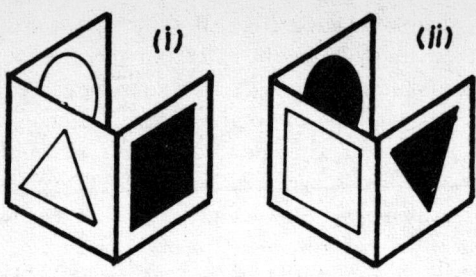

Fig. 29

45 *Symboliosis*

By B and E we have essentially an arrangement such as is shown in Fig. 29(i). Since this is U-shaped, then the three other sides must also be U-shaped. Thus by C and A this other section must be arranged as in Fig. 29(ii). Does D give any clue as to how these two U-shaped sections should be fitted together?

46 *Royal Dilemma*

The princess must try to think of some statement which can be neither true nor false, or which simultaneously contradicts both the king's *forecasts* as to her matrimonial future. What if *she* should make a forecast about who her future husband will be?

47 *Age Old Question*

In the light of M's statements, can you place an upper limit on P's age? Does anything else that was said reduce this maximum? Remembering that no statement is true, can you place a lower limit on P's age? Do the limits you have imposed on P's age now shed further light on statements about M's age?

48 *Amphibolia*

Consider the statements of speakers 3 and 4, in the light of how many people I *know* to be at the bar. Can they be of different classes? What then do you make of 2, and the fact that he disagrees with 1?

49 *But Which Of Us?*

Number the six statements made by the Amphibolians 1–6. Statement 4 cannot be made by a lying P, so X is F or J. If he is a truthful F, what would statements 2 and 5 (taken together) tell you about Y? What would 6 then tell you about Z? Having decided, in the light of all this, what X is, consider statement 4, then see if 2 and 5 now throw still more light on the matter. Then combine 3 with 6.

50 *Tell Me Truly*

Assume A = F (i.e. is truthful), then B = J and C = P, but both B's statements would then be lies, which is impossible for a J. So A < > F. What then can you say about C's opinion of A? If A = J, what of C? If A = P, what of C?

THE SOLUTIONS

In virtually every case, the Solution is dependent upon, and should therefore be read in conjunction with its Lead.

Reasoning given in a Lead is not iterated in its corresponding Solution.

Solutions employ the same mathematical and reasoning conventions as those used in the Leads, and these are explained on page 77.

Each solution begins with an exposition; the actual *Answer* being at the end. This is done so that, should a solver wish to wrestle further with a problem, he can conceal the detailed answer from himself until such time as he wishes to check his final result.

1 *A Little Choosey*
After completing the Lead you should have rows showing:

B	1	–	1	1
L	2	2	2	–

This leaves only 1 to be made up in each of the last three columns if they are each to sum to 5, and these 1s can be accommodated only in the Nothing row, so in order to balance the books, the R and Nothing rows must read:

R	0	0	0	0
Nothing	1	1	1	1
TOTALS	11	5	5	5

Answer: There are 11 Little Chewseys, one eating all three vegetables, three eating cabbage and carrots, one eating cabbage and cauliflower, two eating carrots and cauliflower, one eating cabbage only, two eating cauliflower only, and one very choosey little Chewsey who eschews all this good, wholesome food.

2 Poulterer's Dozen

Multiply both sides of the equation in the Lead by d, and we have the quadratic: $50d - d^2 = 600$, or $d^2 - 50d = -600$. Add *the square of half the coefficient of d* to both sides to get $d^2 - 50d + 625 = 25$, and take the square root of both sides to get $d - 25 = 5$ (making $d = 30$). Like all quadratics, the equation has a second root $25 - d = 5$, but this would make eggs 20p per dozen, and 13 would then cost $21\frac{2}{3}$ pence, which could not have been profferred by Quaddle exactly.

Answer: The eggs were 30 pence a dozen. At this price ($2\frac{1}{2}$ pence each) 50 pence would buy 20 eggs, and 20 pence is the change you would get from a 50p piece were you to buy a dozen eggs. Quaddle's 13 would have cost him $32\frac{1}{2}$ pence.

3 Mix Up

If either O or B are left constant, the quantities F,S,E must be reduced, which is impossible.

If any ingredient *can* remain constant, it must be one whose proportion *falls* (i.e. can be diluted by *other* additions). F and S are the only two whose proportions fall. If F is to remain constant, then the prescribed F:S proportion in b (9:6) will, in column e (the proposed mix) have to become $10:6\frac{2}{3}$, but the 8S we already have cannot be reduced to $6\frac{2}{3}$.

Therefore only S can remain unchanged, making S in column e = 8. Since this is $\frac{1}{3}$ above the prescribed recipe, all other ingredients will have to be increased to $\frac{1}{3}$ above that recipe.

(Whether an egg weighs a gram or a ton, I would still have to add at least 2 eggs to make their number equal to the number of ounces of sugar I already have – one of the proportions prescribed by the recipe. Even a fractional increase in egg-content would increase all other ingredients excessively).

Answer: Add 2oz Flour, $4\frac{1}{3}$oz Oatmeal, $4\frac{2}{3}$oz Butter, and 2 eggs. This makes a mixture of $12:13\frac{1}{3}:10\frac{2}{3}:8:8w$, which represents the same proportions as the prescribed 9,10,8,6,6w.

4 *Measure for Measure*

If you extend the Lead table as suggested, it should finally look like this:

STEP	JUG	FLASK	BOWL
2	0	$f/2 - 8 + j/2$	8
3	j	$f/2 - 8 + j/2$	$8 - j$
4	0	$f/2 - 8 + 3j/2$	$8 - j$

Now F and B finished with 5 pints in each, so $8 - j = 5$, therefore $j = 3$. Substitute this in the FLASK equation and we have $f/2 - 8 + 4\frac{1}{2} = 5$, so $f = 17$.

Answer: Jug capacity 5 pints. Flask 17 pints, and at the outset they contained $1\frac{1}{2}$ pints and $8\frac{1}{2}$ pints respectively.

5 *Well Done!*

Answer: The Mabey solution completes the task by $10\frac{1}{4}$ minutes past 6. Calling the slices A,B,C, and distinguishing their sides by (1) and (2), here is a possible schedule:

6.00   Season A(1)
6.01   Insert A(1), and Season B(1)
6.02   Insert B(1), and Season C(1)
6.04   Remove A(1), Insert C(1), and Season A(2)

6.05   Remove B(1), Reinsert A(1), and Season B(2)
6.06   Turn A to A(2), Remove C(1), Insert B(2), and Season C(2)
6.07½   Remove B(2), Insert C(2)
6.09   Remove A(2), Reinsert B(2)
6.09½   Remove C(2), and Reinsert A(2)
6.10½   Remove A(2) and B(2)

Serve with chips, tossed salad, and a bottle of 1970 *Chateau Margaux*. You've deserved it.

6 A Piece of Cake

Answer: As in Fig. 30, first take a horizontal slice of the cake, one-fifth of the way down from the top. With two further cuts, divide the slice into quarters. Stack the four quarters together to form the second tier, and there you have it – easier to assemble than the pieces proposed in the Lead, with their awkward, crumby, sharp angles.

Fig. 30

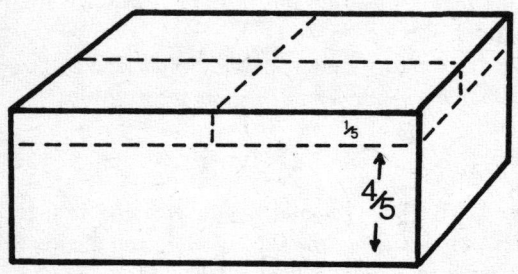

7 Far Enough

At the first question, we must have travelled $\frac{1}{3}$ of the journey, and at the second $\frac{2}{5}$. Now $\frac{2}{5} - \frac{1}{3} = \frac{1}{15}$, and this we were told was half a mile. The whole journey (15/15ths) must then have been:

Answer: Seven and a half miles.

8 *Local Colour*

Like Area 2, Area 4 must be Y or G, and since they touch one another, they cannot both be the same colour. Since 6 touches both 2 and 4, it must therefore touch a Y *and* a G, as well as Area 3 which is B. 6 then must be R.

Similarly, by considering 14,13,15, it can be shown that 12 must be G. Then since 1 touches 12 it cannot be G, neither can it be B or R (because it touches 3 and 6). Therefore 1 is Y, which means that 2 must be G, and 4 is Y. Then 7 (touching 1,6,12) must be B. Since 8 touches 2,6,7, it must be Y. Since 10 touches 8,7,12, it must be R. Since 13 touches 10,12,14, it must be B, making 15 R. Since 9 touches 2,8,10, it must be B. Since 11 touches 10,13,16, it must be Y.

Answer: Fig. 31 shows the full colouring.

Fig. 31

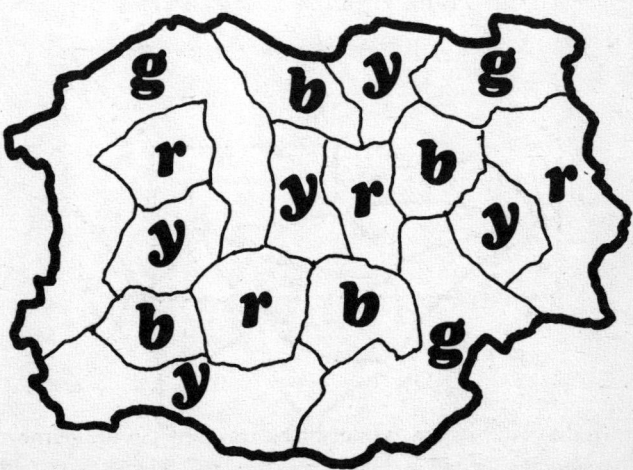

9 Mystery Tour

Following the Lead we see that in order to reach Town 10, Tour X must have the cyclic order 1–2–6–10–11–7–3–1.

Since 7 and 11 are the only two towns common to both tours, one must be B and the other L (these being the only two letters common to the bureau's itinerary for these two tours). Consider the remaining six-link tour. It offers B–J direct to a town on Tour X by an alternative route, but if B=7, B could not offer an alternative direct route to J, which (being two steps from B) would in this case be 2,10,5 or 12. Therefore B<>7.

Therefore B=11 making J=6 or 8.

Assume J=6, then Tour X would be:

$$B - K - J - T - C - A - L - B$$
$$11 \quad 10 \quad 6 \quad 2 \quad 1 \quad 3 \quad 7 \quad 11$$

but our six-link tour offers J–C direct, a link which is unavailable in this case. Therefore J<>6, so J=8.

Then Tour X is the Highland Tour which must be:

Fig. 32

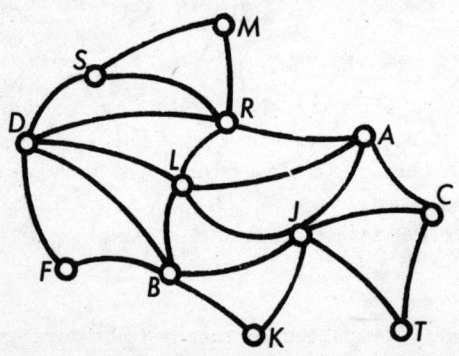

$$B - K - J - T - C - A - L - B$$
$$11 \quad 12 \quad 8 \quad 9 \quad 5 \quad 4 \quad 7 \quad 11$$

The short Scenic tour must then be:

$$B - J - C - A - R - D - B$$
$$11 \quad 8 \quad 5 \quad 4 \quad 3 \quad 6 \quad 11$$

Making the Lowland Tour:

$$B - L - R - M - S - D - F - B$$
$$11 \quad 7 \quad 3 \quad 1 \quad 2 \quad 6 \quad 10 \quad 11$$

Answer: Fig. 32 shows the completed map.

10 *Air Miles*

Since PT + TA = PA, PTA must be a straight line. Since TA,MT,MA are in the proportion 3:4:5, ATM must be a right-angled triangle with the right-angle at T. Then MTP is also a right-angled triangle (again of 3:4:5 proportion). Therefore:

Answer: The distance M to P is 200 miles.

11 *Touranian Tours*

In Fig. 26, both the triangles ABC and CED are isosceles right-angled triangles, and since AC = DC the triangles are congruent. The side AB of the triangle ABC therefore equals the side CE of the triangle CED.

Answer: It is 20 miles from A to B.

12 *Heare and Thare*

The *apparent* Normal Service time is 8.50 for the outward journey and 2.50 for the home journey, an apparent saving of

6 hours. Crossing the same zone the Supersonic flight will enjoy the same apparent 6-hour saving, cutting its apparent outward time from 5.50 to -10 minutes.

Answer: The Supersonic flight arrives back at Heare at 22.40, i.e. ten minutes *before* its local time of departure from Thare.

13 World Tour

Answer: Here are enough countries and islands to give you a fairly good round-the-world tour; the twelve you really should have got being given in italics: Adi, *Chad*, *China*, Dalma, Hainan, *Haiti*, *India*, Kanai, Lanai, *Laos*, Linosa, *Mali*, Man, Moa, Mon, Mona, Oland, *Oman*, Salina, *Samoa*, Samos, Sanana, Sanda, *Solomons*, *Somalia* (or Somaliland), *Thailand*, Thasos, Tilos, Tina, *Trinidad*.

14 Red or Black

The Lead equation $x = 5y - 26$ tells you that y cannot be less than 6 (otherwise you couldn't subtract 26 from 5y). The suggested second equation will yield $z = 52 - 8y$, which means that y cannot be more than 6. If it can be neither more than nor less than 6, it must be 6. From our first equation above this makes $x = 4$, and from the second, $z = 4$. The rest is easy.

Answer: (Pile 1) 12 Blacks and 4 Reds. (Pile 2) 6 Blacks and 18 Reds. (Pile 3) 8 Blacks and 4 Reds.

15 Think of Two Cards

Step 3 will give $12S + 2H$; Step 4 $11S + 2H$; Step 5 $11S + H$. Now, 11S will divide by 11 (without remainder). Also (since H must be less than 11, and therefore not divisible by 11) H will appear as the remainder when $11S + H$ is divided by 11.

Answer: Divide your victim's answer by 11. The quotient will be the value of the Spade, and the remainder will be the value of the Heart.

16 Straight Answer Needed

We have already shown that $5 <> A$. Now:

By (ii) $5 <> T$; $4 <> A$.

By (vi) $5 <> J$; $1 <> J$.

By (iii) $5 <> Q$. Therefore we are left with $5 = K$.

By (v) $A <> 3$. Therefore, since we already know $A <> 5$ and $A <> 4$, $A = 1$ or 2. Therefore by (v) $c = 4$ or 5.

Now if $J = 2$, then by (vi) $3,1 = s$, then by (iv) d,h must occupy 4,5, but c occupies one of these places. So $J <> 2$.

If $J = 3$, then by (vi) 4,2 are s, leaving no place (according to iv) for d and h. Therefore $J <> 3$, nor (as we have already proved) can J be 2,5 or 1. Therefore $J = 4$, making by (vi) $5,3 = s$. Therefore $c = 4$, making by (v) $A = 1$, and by (iii) $T = 3$ and $Q = 2$. From (iv) $2 = h$ and $1 = d$.

Answer: (1) Ace of Diamonds, (2) Queen of Hearts, (3) Ten of Spades, (4) Jack of Clubs, (5) King of Spades.

17 Handy Question

If $A = 2$, we should have a minimum of 2A, 3h, 4Q, 5s, 6c, i.e. $3 + 5 + 6 = 14$ *suit*-named cards, which is impossible. Therefore $A = 1$.

Since there are 4X, each of the four suits is represented, so there must be at least 1 trump. Since $J > t$, then $J = > 2$.

With (from the story) at least 2F, we have at least 4X, 3Q, 2J, 2F, 1A, accounting for 12 cards, leaving the possibility of only 1K.

Since $Q = 3 > h > A (= 1)$, there must be 2h. Since $J (= 2) > t$, there can be only 1t, and it must be the d, because both c and $s > h$.

If $s = > 5$, there would be at least 6c, making 14 cards, so $s < 5$, but $s > Q (= 3)$, so $s = 4$. With 2h and 1d, we must have 6c to make up the 13. With all this decided, the rest is easy.

Answer: A,K,Q,J,5,2 of Clubs; Q,J,5,2 of Spades; Q,2 of Hearts; 2 of Diamonds (which were Trumps).

18 *Win or Lose*

Since T started with $\frac{1}{5}$ of the total pool, and finished with $\frac{1}{4}$, he won $\frac{1}{4}-\frac{1}{5}=\frac{1}{20}$. Similarly J won $\frac{1}{3}-\frac{1}{4}=\frac{1}{12}$, so together they won $\frac{2}{15}$. So B must have *lost* $\frac{2}{15}$ which, we are told, was £2. Therefore the pool was £15. T's finishing total is $\frac{1}{4}$ of this, J's $\frac{1}{3}$, and B's the remainder. Hence:

Answer: Tony finished with £3.75, Jim with £5, Bob with £6.25 (having started with £3, £3.75, £8.25 respectively).

19 *Pick-a-Pack*

It follows from the Lead that, if your opponent can leave a SAFE combination then HE wins. Now look at the full, seven-card array in Fig. 27. It is YOUR turn, so a(cd)Q-HE, b(de)R-HE, c(ad)Q-HE, **dT-YOU**, e(c)S-HE, f(cd)Q-HE, g(ad)Q-HE, ad(c)Q-HE, bf(cde)P-HE, eg(a)Q-HE, ab(cde)P-HE, cd(a)Q-HE, **ceS-YOU**, de(b)R-HE. Notice that only d or ce *immediately* produces a SAFE combination (T or S) for you to leave.

Answer: Claim either (1) the Ace of Hearts, or (2) the Ace of Clubs and Ace of Spades.

20 *Very Tricky Question*

Assume T12 or T13 contains a 5, then by (m) 11 must contain 5, so by (j) T8 and T10 would contain 5s, but there would be no 5 for T9 to make all 5s fall consecutively. So T12 and T13 do not have 5s, so the 5s are in Tricks 8,9,10,11.

Moreover, all 10s are accounted for, and no 5s are available for T1,2,3, so T3(SL) = < 9, so by (c) and (e) T1,2,3 become:

(1)	2	3	4	6
(2)		6	7	8
(3)	8	9		

Now Tricks 1,2,6,7 contain no picture cards, neither by (g) does T5, so the remaining eight hands contain all 16 picture cards, so by (b) each contains exactly two of them. Consider the picture cards in T4, T12, T13.

Taking into account (f) and (l) we can show:

$$T4(H) > T4(SH) = T12(H) > T12(SH) = T13(H) > T13(SH)$$
$$A \quad\quad K \quad\quad K \quad\quad Q \quad\quad Q \quad\quad J$$

Since two Ks and two Qs are now accounted for, the two picture cards of (the identical) Tricks 8,10,11 must be As and Js, leaving two Ks and two Qs to fill the vacancies in 3 and 9.

Since two of every value (except 5) have been assigned, the three identical Tricks 8,10,11 must each have a 5, so by m, T9 must have 5.

By (k) we can show that T9(SL) and T5(SL) are 7, forcing 8 and 9 to complete the top end of T5. You should now be able to assign the only eight cards remaining (9,6,4,4,3,3,2,2).

Answer: 1st 6,4,3,2. 2nd 8,7,6,4. 3rd K,Q,9,8. 4th A,K,3,2. 5th 9,8,7,4. 6th 6,4,3,2. 7th 10,9,8,7. 8th A,J,10,5. 9th K,Q,7,5. 10th A,J,10,5. 11th A,J,10,5. 12th K,Q,3,2. 13th Q,J,9,6.

21 *What's in a Name?*

Answer: Apart from Bert (mentioned in the puzzle) and Mandy (in the Lead) other possibilities are: Andy, Nesta, Ann, Anne, Di, Diana, Al, Fred, Freda, Dan, Gus, Sam, Myra, Helen, Len, Ida, Ada, Mabel, Abe, Elsa, Eve, Ron, Carol. (Al, Abe, and Di are abbreviations for Albert, Abraham, and Dinah.)

And it was of course thoughtful of Judy to include Dad and Ma.

22 *French and English*

Consider a call to Room 3. By the story, it does not hold the Legrands. If an English male or French female should answer, it cannot be the Dubois, so it must be the Smiths, and since the Dubois cannot be in 2, they must be in 1, leaving 2 to the Legrands. A call to Rooms 1 or 2 will prove inconclusive should a French voice answer. Such reasoning leads us to:

Answer: One call to Room 3 will suffice: If an English male or French female should answer, the occupants of the rooms are (1) Dubois, (2) Legrand, (3) Smith. If a French male or English female should answer, then (1) Legrand, (2) Smith, (3) Dubois.

23 *Tabulation*

By 3 and 8, (ii) = E. By 5, (iii) = h. By 5, (iv) = p. By 6 and 8, (vi) = D. By 8, (iii) = B. By 3, (ii) = k. By 8, (v) = n.

Answer: Starting from Georgina Aintwright's right, the ladies are: Janet Darfter, Norma Fewtile, Patricia Cockshure, Helen Banmen, Katherine Ennycawse.

24 *Smoko*

Nor can the cellist be S (who smokes) nor Q (who stands) and is therefore T. So the pianist is not T, nor Q (who stands) nor R (who no longer smokes) but is S. R is not the violinist (who asked him for a light) so must be the conductor, making Q the (standing) violinist.

Answer: Quentin is the violinist, Raymond the conductor, Simon the pianist, Thomas the cellist. The only major Beethoven work calling for a *standing* (i.e. solo) violinist *and* a pianist is the *Concerto for Piano, Violin, Cello and Orchestra, Opus 56*.

25 Sorting out the Mess

Since b+d = Brig+Cav, and d = Lt, then b = Brig, d = Cav. Since b+e = Art+Inf, then a = Eng, and d = Cav, then (by elimination) c = Sig.

Since a+e = Maj+R, and a = T, then e = R, and a = Maj, leaving d = Q. Since d+e = Q+Inf, and d = Cav, then e = Inf, leaving b = Art. Since Col < > Sig, then c < > Col, leaving e = Col, leaving c = Capt. Therefore since the Captain (now known to be c) is not the Beer-drinker, d = Beer. Since Gin won more than Brandy, a = Gin, leaving e = Brandy.

Answer: Brigadier Simpson (Artillery) won £6, and drank Whisky. Colonel Rogers (Infantry) won £1, and drank Brandy. Major Tait (Engineers) won £9, and drank Gin. Captain Powell (Signals) won £5, and drank Whisky. Lieutenant Quist (Cavalry) won £3, and drank Beer.

26 Gooseberry Fool

The mathematics of the story is impossible unless it is assumed that George is a woman, and Evelyn is a man. There are two very well known such persons who 'went on writing':

Answer: George Eliot, born 1819, married (John Cross) 1880, and died in the same year. Her best known work is *Middlemarch*, published 1871. Evelyn Waugh (pronounced 'War'), born 1903, died 1966. His best known work is *Decline and Fall*, published 1928.

27 Catch Patience

From the Lead, the first seven catches must be 12,10,4,11,6,3,5. From 5 one can reach only 1 or 9. If 5,1, then ,5,1,7,2,8, with no prospect of reaching 9. Therefore we must have 5–9, giving ,5,9,7,2,8,1. Add this to the first sequence.

Answer: 12,10,4,11,6,3,5,9,7,2,8,1.

28 Odds and Evens

If you claim 3, he can take 3 to make up his final odd total. If you claim 1, he can take 2; then, whatever you claim, he can make up his final odd total on his next claim.

Answer: Claim 2 matches. If your opponent replies by taking 1 or 2, you take the remaining 3. If his first reply is to take 3, you claim 1.

29 Grand Chain

With the biggest tiles accounted for, *a*-group tiles must bear 9′9′7 or 9′8′8 dots. Of the tiles bearing these numbers of dots, the only ones bearing a 3 to open the group are 3–6;3–5;3–4. 3–4 would force 3–4, 4–5. 3–6 would force 3–6, 6–2. In neither case is any remaining tile left to follow on. So the first tile is 3–5, leading to 3–5, 5–4, 4–4, *b*.

Now *b*-group totals 18, needing four tiles bearing 4′4′5′5 dots. Among such tiles remaining (5–0;4–1;3–2;3–1;2–2) only 4–1 has a 4 to link with *a*-group, so we must have *a*, 4–1, 1–3, 3–2, 2–2, *c*.

Now *e*-group totalling 18 must be of the form ?–5 (to link with *f*-group). Possible tiles for *e*-group are those bearing 6,6,6 or 5,6,7 dots, but the 6-dot tiles which remain have no *internal* links between themselves; so we must accept the 5,6,7 batch; and since 5–0 is the only remaining tile totalling 5, we must use it as the end tile of the group, leading to *d*, 1–6, 6–0, 0–5, *f*.

The only remaining tile now showing a 1 is the 5–1, so it must end *d*-group to link with *e*-group, and must be preceded by the 2–5. To make the group total 21, we must then have *c*, 6–2, 2–5, 5–1, *e*.

Only the 6–3;4–3;4–2;3–3 remain for *c*-group, and to link with both *b*-group and *d*-group, we must start with 2 and end with 6. The only possibility is *b*, 2–4, 4–3, 3–3, 3–6, *d*.

Answer: 3–5, 5–4, 4–4, 4–1, 1–3, 3–2, 2–2, 2–4, 4–3, 3–3,

3–6, 6–2, 2–5, 5–1, 1–6, 6–0, 0–5, 5–5, 5–6, 6–6, 6–4, 4–0, 0–0, 0–2, 2–1, 1–1, 1–0, 0–3.

30 *All Buttoned Up*

Any initial choice other than those given below will permit a canny opponent to reduce the saucers to a SAFE combination himself, and let him jump in on the winning strategy.
Answer: In the two-saucer game, draw 1 button from the saucer holding 8. Thereafter aim for the lowest SAFE combination listed in the Lead.

In the three-saucer game, draw 1 button from the saucer holding 3. Thereafter again aim for a lower SAFE combination.

31 *Diamond Chain*

If White claims 5, Black 18, White 4, then Black could claim 7. He then has forks on 2;3, 8;12, 22;23. Whichever of these White claims, Black can, on each move, claim the other one of the pair, and complete his chain. Whichever way we look at it, 18 becomes a vital pivot for Black, and White must block him on this circle immediately.

If White claims 18, the best Black can then do is to claim 15, for if he does not, White can claim either 14 (forking on 10;15) or 19 (forking on 15;20) as well as having a fork on 17;22. If Black does claim 15, White replies with 24 (forking on 20;25) with further forks at 17;22, 22;23, 19;23.
Answer: Claim circle 18.

32 *Checkel*

Here are possible positions for you, with your opponent's winning responses in brackets: CDI, DEJ, EIJ, DIJ, DEI (all answerable by KLQ), FKL(DEJ), IJO(KLQ), KLQ(DEJ), PKL, PQV, QUV, QVW, PUV, PQU (all answerable by DEJ), TYX(KLQ), LMQ(DEJ). That

doesn't leave you much room for manoeuvre, does it?
Answer: Break up that dangerous bottom-left corner by claiming KPQ or LPQ. Your opponent cannot then avoid killing one of the remaining corners, leaving you to take the other.

33 *Break'n the Bank*

Answer: Make your first three taps on *any* three cards. Thereafter make your taps to spell out the word BREAK'N. Your final tap must then always be resting on the card your victim has chosen when he says 'Change Please!'

34 *Take Your Pick*

The only other 15+11 start leaves 4+7 or 2+9, yielding a grand total of only 37, which is worse than the best in the Lead. Only one other combination surpasses 39.
Answer: The best possible selection is 14,12,9,5, totalling 40.

35 *Square Pairs*

(1–16) Following on from the Lead, 14 can be paired only with 11, 5 with 4, 12 with 13, 8 with 1, 15 with 10, leaving 6 with 3.

(1–18) Clearly 16 can be paired only with 9, 17 with 8, 18 with 7. Then 2 can be paired only with 14, 11 with 5, 4 with 12, 13 with 3, 1 with 15, leaving 6 with 10.

(1–20) Clearly 18 can be paired only with 7. Then 9 with 16, 20 with 5, 11 with 14, which leaves 2 crying 'wee wee wee' all the way home.
Answer: (1–16) 1+8, 2+7, 3+6, 4+5, 9+16, 10+15, 11+14, 12+13. (1–18) 1+15, 2+14, 3+13, 4+12, 5+11, 6+10, 7+18, 8+17, 9+16. (1–20) You were only asked to try your luck, and your luck didn't hold. The task is impossible.

36 Each Way Accumulator

Assume ghi = 819. Because h = 1, there is a carry from b + e, so a + d = 7, so a'd = 2'5 or 3'4. If 2'5, then of the remaining 3,4,6,7, 3'6 must be allocated to c'f which must together total 9. Then a + b = c must be either 5 + 7 = ?2 or 2 + 4 = 6, but in the former case c = 2, whereas it should be 3;6. If a + b = c is 2 + 4 = 6, then to add up the centre line would have to be 573, but when adding across 852 + 174 < > 936. On the other assumption that a'd = 3'4, then 2,5,6,7 remain, of which 2'7 must be allocated to c'f (totalling 9). Then abc would be 3 + 5 = 8, or 3 + 6 = 9, or 4 + 5 = 9, or 4 + 6 = ?0, making c 8;9;0, whereas in this case c should be 2;7. Therefore ghi < > 819.

Assume ghi = 729. Then since h = 2, there is a carry from b + e, so a + d = 6, so a'd = 1'5, leaving 3,4,6,8 of which 3'6 must be allocated to c'f. Then if a + b = c is 1 + 4 = 5, or 1 + 8 = 9, or 5 + 4 = 9, c would be 5;9, whereas it should be 3;6. But if a + b = c is 5 + 8 = ?3 we have 583 + 146 = 729, which twisted becomes 715 + 248 = 963, WHICH IS A SOLUTION.

Assume ghi = 369. Then a'd = 1'2 making b'e = > 4'5, giving a carry, thus invalidating 1 + 2 = 3 in the first column. So ghi < > 369.

Assume ghi = 639. Then a'd = 1'5, 2'4, or 1'4 (+ carry). If 1'5 or 2'4, there will be no carry from b + e, so b + e = < 3, and no numbers for b,e can exist (for either 1 or 2 is being used in the combinations being considered). Therefore a'd = 1'4(+ carry), leaving 2,5,7,8, of which 2'7 must be allocated to c'f. Then if a + b = c is 1 + 5 = 6, or 1 + 8 = 9, or 4 + 5 = 9, c would be 6;9, whereas c should be 2;7. But if a + b = c is 4 + 8 = ?2, we have 482 + 157 = 639, which twisted becomes 614 + 358 = 972, WHICH IS A SOLUTION.

Assume ghi = 459. Then a'd = 1'3 or 1'2(+ carry). If

a'd = 1'3 then b + e = < 5, making b'e = 1'4, 1'3, 1'2. In each case, 1 is used in a'd, and is therefore not available. If a'd = 1'2(+carry) we are left with 3,6,7,8, of which 3'6 must be allocated to c'f to equal 9. Then if a + b = c is 1 + 7 = 8, or 1 + 8 = 9, or 2 + 7 = 9, or 2 + 8 = ?0, c would be 8;9;0, whereas c should be 3;6. So ghi < > 459.

Assume ghi = 549. Then a'd = 2'3 or 1'3(+carry). If a'd = 2'3, then b + e = < 4, making b'e = 1'3 or 1'2. In each case one of the pair is used in a'd, and is unavailable. If a'd = 1'3(+carry) we are left with 2,5,7,8, of which 2'7 must be allocated to c'f to equal 9. Then a + b = c must be 1 + 6 = 7 or 1 + 8 = 9 or 3 + 6 = 9 or 3 + 8 = ?1. In the first case 167 + 382 = 549, but 531 + 486 < > 927. In the other cases, c would be 9;1, whereas c should be 2;7. Therefore ghi < > 549.

Answer: All cases having been now examined, only two possibilities have emerged:

```
5 8 3        4 8 2
1 4 6   or   1 5 7
7 2 9        6 3 9
```

37 *Crossnumber*

From (5), d(dn) becomes 5746. From (3), 9?? + 81 = 10?6, so the last digit of c(dn) must be 5 and from (2) its middle digit is 2. Then by (3) i(ac) = 1006.

Answer: ACROSS: a = 169, e = 34, f = 27, g = 80, h = 54, i = 1006.

DOWN: a = 13, b = 6400, c = 925, d = 5746, g = 81.

38 *PSC Square*

Turning to array (y), all combinations of p and q can be

similarly eliminated with the exception of those with rows a,e,f,g reading 5832, 3481, 8649, ?447. For a(dn) the only primes taking the form 538?, are 5381 and 5387, but the latter would make g(ac) 7447 (which is composite). The only possibility is for a(dn) to read 5381.

Answer:

5	8	3	2
3	4	8	1
8	6	4	9
1	4	4	7

39 *Digi-Tally*

To total 17, the last two of 3dn must be 7'9. If the last were 9, 10ac would need one zero, but both first and last digits of 10ac start a number; therefore *3dn = 197*. Since 8ac now contains a 9, the first cannot be > 8, so to attain 21 its centre digit is not < 4. But it cannot be > 4 (because 2dn, totalling only 10, can be only 0'1'2'3'4). So *8ac = 849*.

Now 1dn can be only 9'8'7'4 or 9'8'6'5. If the latter is used, the smallest digit assignable to 1ac is 5, but this would make 1ac = 50, and 2dn would start with zero. So we must accept *1ac = 41*.

10ac must now be 127 or 271, but cannot be the former since 1 already occurs in 2dn, so *10ac = 271*.

Of the digits in 1dn, only 9 or 7 remain to start 14ac (totalling 9) but 900 would use two zeros, so it must be 7,2'0, but 2 already appears in 2dn, so *14ac = 702*, and *1dn = 4897*. The second digit of 17ac must be the remaining 3 of 2dn. 17dn (with no zeros as starters) can have no digit > 5, and this digit will be needed now to let 17ac reach 17, making *17ac = 539*.

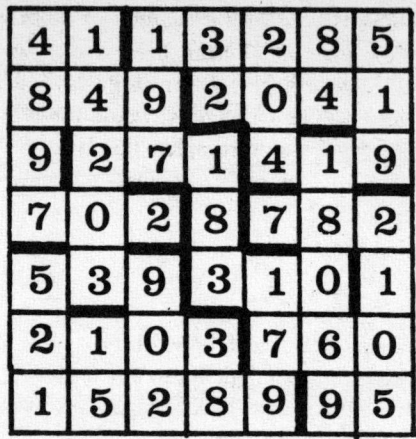

Fig. 33

Now 21ac must be 0′1′2′3, but since zero cannot start a number, it must be the 3rd digit, making *18dn = 902*. 17dn can be only 512 or 521, but the former would give two 2s in 25ac, so *17dn = 521*.

25ac now becomes 1?2??, the missing digits being 9′7′6 or 9′8′5, but the first is too large because 22dn totals only 6. So 25ac has 9′8′5, the 5 going to make *22dn = 15*. Then *21ac = 2103, 23dn = 38, 25ac = 15289*.

Now 19ac is 310 or 301, but the latter would give two 1s in 13dn, so *19ac = 310*, making *20dn = 179* and *11dn = 183*. 13dn is now 1?0??, so the missing digits are 9′8′6, forcing *24ac = 760*.

The remaining digits of 16dn totalling 8 are then 1′2′5 or 1′3′4. To make 26ac = 14, we need the 5 from 16dn, so *26ac = 95*. This makes *13dn = 18069*. The first digit of 16dn, i.e. the last of 15ac, is now either 1 or 2. If 1, then 15ac would need two 8s, so *15ac = 782, 16dn = 2105*. And that's that!

Answer: The only possible answer is that shown in Fig. 33.

40 Exhibitionism

The complete pile would have been $6 \times 6 \times 4 = 144$ half-blocks (i.e. 72 whole blocks). It can easily be visualised that 4 *double*-blocks (each $12 \times 12 \times 6$cm) would fill all but the top layer, and another 2 would complete that layer. So we need 6 *double*-blocks. That means 12 ordinary blocks are missing from the original 72.

Answer: 60 blocks remain.

41 Cut It Out

11 six-square shapes can be folded into a cube: therefore a cube may be cut and unfolded into any of these 11 shapes.

Answer: The 11 distinguishable shapes into which a cube may be cut are shown in Fig. 34.

Fig. 34

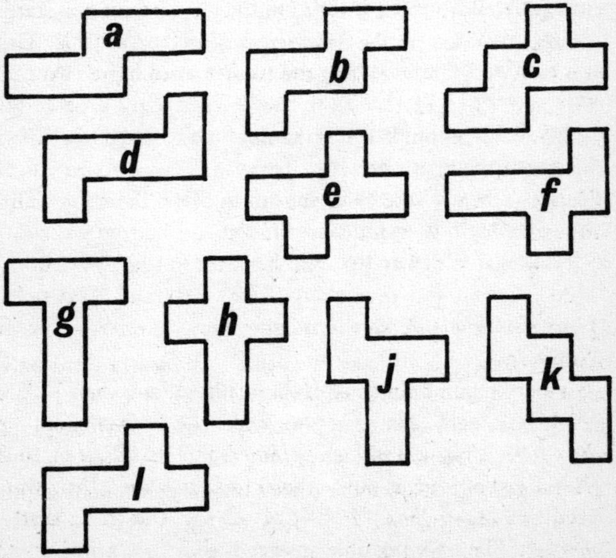

Fig. 35

42 Make It Up

The results of trying to cover 23 in Fig. 28(iv) with a,f,j,g are: a isolates two squares, f isolates three, j isolates two, g isolates both corners. Therefore only the cross-shaped h can cover 23 (taking in 19,13,14,15,9 as it goes). Then only k or c can reach 5, but k would isolate a corner. So c covers 5, leaving b to complete the pattern.

Answer: The arrangement shown in Fig. 35 is basically unique (though, by symmetry, the top and bottom pairs can each be independently reflected).

43 In Black and White

Systems a,g (with all faces the same) can clearly yield only one way. So do b,f (with only one odd face). In c the two Ws may be either adjacent or opposite, so there are two ways for c. Similarly there are two ways for e. In d, the three Bs may be arranged either around a corner or in a continuous band, so d also yields two ways. So a,b,c,d,e,f,g yield respectively 1,1,2,2,2,1,1 possibilities.

Answer: There are ten distinguishably different ways of colouring the cube black and white.

44 Rubikitis

Whatever colours appear on top and bottom, the cube may be turned until one of the 'equator' sides faces you. Call this side p. Then the remaining three equatorial faces may (walking around the cube in a clockwise direction) be arranged qrs, qsr, rqs, rsq, sqr, srq – giving six distinguishable combinations for each and every one of the five top/bottom arrangements, i.e. 5×6 combinations all told.

Answer: There are 30 distinguishably different pristine Rubik cubes.

45 Symboliosis

From D we see that the apex of the black triangle must point towards the black square, so Figs. 29 (i) and (ii) must be fitted together as shown in Fig. 36. By orientating such a cube to match the five pictures in the problem, we can read off the top symbol of each.

Answer: The top symbol in each case is: (A) white circle, (B) black circle, (C) white circle, (D) white triangle, (E) black triangle.

Fig. 36

46 Royal Dilemma

Human ingenuity being what it is, there is virtually no limit to how many ways the princess can contradict herself with a statement that is neither true nor false, e.g. 'The statement I am now making is false,' but the king may interpret this as treasonably irrelevant, and claim that she has lost her chance to influence his decision. She would be better advised to make a wholly submissive, but impossible, choice of husbands:

Answer: She should say, 'I shall marry none other than Grewsumm.' If then the king were to marry her off to Grewsumm, she would have spoken the truth and could be claimed by Pompuss. If the king married her off to Pompuss, she would have lied, and could be claimed by Grewsumm. Since she can be claimed by both, the king can marry her to neither.

47 Age Old Question

When M says 'P is at least 39,' we know that P cannot be more than 38. But she is not 38 because that is the age P claims to be. So P is less than 38, but must be over 36, otherwise M (whom P falsely claims to be 5 years older) would not be more than 40 (whereas we know that her claim to be not more than 40 is false). So since P is more than 36, but less than 38, she must be 37, and M must be 41 (if both women's statements about M's age are lies).

Answer: Penelope is 37. Margaret is 41.

48 Amphibolia

Since there are five Amphibolians, 3's statement about two Fellahs implies there are three Phibbas. 3 and 4 therefore agree, and must belong to the same class. If both are Fs they are the *only* two Fs, making 2's statement truthful, which means that he *also* is an F, making three Fs, which (according

to our assumption) is impossible. Thus 3 and 4 must both be Ps. Now 1 and 2 disagree, so one of them must be a P, making (with 3 and 4) at least 3 Ps. Therefore 2 tells the truth, so 2 = F, and 1 = P. 5 must then also be a P, otherwise 3 and 4 would be truthful Phibbas, which would be absurd.
Answer: The second speaker was a Fellah. All the others were Phibbas.

49 *But Which Of Us?*

Following on the Lead, *assume* X = F, then there *is* one J, so (by 2) Y < > P, and (by 5) Y < > F, making Y = J. But by 3, Z < > F, and (by 6) Z < > P, making Z yet another J, whereas X said there was only one J. *So* X < > F. So X must be J, making 4 a lie, which means that X's first statement is the truth. Then (in 2 and 5) Y tells the truth, making him an F, while (in 3 and 6) Z lies, making him a P.
Answer: Xzymith (with truth/lie) is a Joker. Yxzmith is a Fellah. Zxymith is a Phibba.

50 *So Tell Me Truly*

By the Lead A < > F, therefore C who asserts this fact cannot be a lying P. Moreover, only two possibilities remain for A. He is either a J or P:

(i) Assume A = J. Then since his first statement would be a lie, his second would be true, i.e. B would be a J. Then since B's first statement would be a lie, his second would be true, i.e. *C would be J*.

(ii) Assume A = P. Then by his second statement B < > J, but B's first statement would be true, so B < > P; therefore B would be a truthful F, in which case his second statement would also be true, i.e. *C would be J*.

Therefore in the only two possible cases (i) and (ii), C = J. Of the two men, A may be a lying Phibba, and B a truthful Fellah; or they could both be unreliable Jokers – we shall never know. But we do know that C is a Joker, and since her first statement about A is (as we found in the Lead) certainly true, her second statement about me must be false.

Answer: Alas, the sweet young thing does not love me. I am devastated! My life is at an end! So is the book! I can write no more! It is finished!

Death Among the Stars

Harry James was familiar with London lodging-housekeepers, falsely smiling, trying to eke out a living of sorts from tenants who loathed them, rack-rented by landlords, everlastingly tormented by antique plumbing and complaints about horsehair mattresses. The woman who ran number fourteen was a cut above the usual run, a pleasant middle-aged blonde who whitened under her make-up as the Inspector explained his errand.

Percy Button had occupied two rooms on the second floor, a bedroom and a sitting-room, both rather chintzy but pleasant and obviously recently decorated. The living-room was lined with bookcases, of the extending type. Harry shook his head. He had never seen so many current reference books, town council reports included.

"A bit of a crease here, Harry," said the Sergeant, for a moment unbending. He addressed the Inspector by his Christian name around four times a year, excluding parties.

The wall-to-wall synthetic carpet showed a bend in the corner of the room. The Sergeant dropped ponderously to his knees.

"The tacks have been removed." The corner of the carpet flipped up and over. Honeybody removed a floorboard. Inside was a leather box.

"Steel underneath and a combination lock." Honeybody caressed it.

"Leave it," said Harry, a memory of one of his police courses flickering in his mind. "We carry it tenderly back to the lab."

Other titles in the Walker British Mystery Series

Peter Alding • MURDER IS SUSPECTED
Peter Alding • RANSOM TOWN
Jeffrey Ashford • SLOW DOWN THE WORLD
Jeffrey Ashford • THREE LAYERS OF GUILT
Pierre Audemars • NOW DEAD IS ANY MAN
Marion Babson • DANGEROUS TO KNOW
Marion Babson • THE LORD MAYOR OF DEATH
Brian Ball • MONTENEGRIN GOLD
Josephine Bell • A QUESTION OF INHERITANCE
Josephine Bell • TREACHERY IN TYPE
Josephine Bell • VICTIM
W. J. Burley • DEATH IN WILLOW PATTERN
W. J. Burley • TO KILL A CAT
Desmond Cory • THE NIGHT HAWK
Desmond Cory • UNDERTOW
John Creasey • THE BARON AND THE UNFINISHED PORTRAIT
John Creasey • HELP FROM THE BARON
John Creasey • THE TOFF AND THE FALLEN ANGELS
John Creasey • TRAP THE BARON
June Drummond • FUNERAL URN
June Drummond • SLOWLY THE POISON
William Haggard • THE NOTCH ON THE KNIFE
William Haggard • THE POISON PEOPLE
William Haggard • TOO MANY ENEMIES
William Haggard • VISA TO LIMBO
William Haggard • YESTERDAY'S ENEMY
Simon Harvester • MOSCOW ROAD
Simon Harvester • ZION ROAD
J. G. Jeffreys • SUICIDE MOST FOUL
J. G. Jeffreys • A WICKED WAY TO DIE
J. G. Jeffreys • THE WILFUL LADY
Elizabeth Lemarchand • CHANGE FOR THE WORSE
Elizabeth Lemarchand • STEP IN THE DARK
Elizabeth Lemarchand • SUDDENLY WHILE GARDENING
Elizabeth Lemarchand • UNHAPPY RETURNS
Laurie Mantell • A MURDER OR THREE
John Sladek • BLACK AURA
John Sladek • INVISIBLE GREEN